VERIFICATION and VALIDATION of RULE-BASED EXPERT SYSTEMS

VERIFICATION and VALIDATION of RULE-BASED EXPERT SYSTEMS

Suzanne Smith, Ph.D.

Converse College
Spartanburg, South Carolina

Abraham Kandel, Ph.D.

Department of Computer Science and Engineering
University of South Florida
Tampa, Florida

CRC Press

Boca Raton   Ann Arbor   London   Tokyo

Library of Congress Cataloging-in-Publication Data

Smith, Suzanne, 1953–
    Verification and validation of rule-based expert systems / Suzanne Smith, Abraham Kandel.
        p.   cm.
    Includes bibliographical references and index.
    ISBN 0-8493-8902-X
    1. Expert systems (Computer science) 2. Computer software--Verification. 3. Computer software--Validation. I. Kandel, Abraham. II. Title.
QA76.76.E95S66 1993
006.3′3—dc20   93-1553
                                                                CIP

    Direct all inquiries to CRC Press, Inc., 2000 Corporate Blvd., N.W., Boca Raton, Florida 33431.

© 1993 by CRC Press, Inc.

International Standard Book Number 0-8493-8902-X

Library of Congress Card Number 93-1553
Printed in the United States of America 1 2 3 4 5 6 7 8 9 0
Printed on acid-free paper

PREFACE

The objectives of this book are to define verification and validation for expert systems, to delineate the activities for comprehensive verification and validation of rule-based expert systems, and to present a comprehensive set of techniques and tools for the verification and validation of rule-based expert systems. Our strategy in accomplishing these objectives is to adapt the concepts of verification and validation in conventional software for use in the verification and validation of rule-based expert systems and to specify new methods, techniques, or tools for verification and validation where the uniqueness of rule-based expert systems demands them. Currently, verification and validation of expert systems are informal, subjective, time-consuming, tedious, and arbitrary in both the creation and execution of test cases. In this book, we present a systematic, comprehensive approach to the verification and validation of rule-based expert systems. Additionally, a complete set of techniques and tools, which provides a more formal, objective, and automated means of verifying and validating rule-based expert systems, is described.

Suzanne Smith
Abraham Kandel

May, 1993

THE AUTHORS

Suzanne Smith, Ph.D., received her B.Sc. in Education from Baylor University, M.A. in Education from Vanderbilt University, M.Sc. in Computer Science from University of Southwestern Louisiana, and Ph.D. in Computer Science from Florida State University. Dr. Smith is currently an Assistant Professor at Converse College in Spartanburg, South Carolina. She has been a software engineer at Lockheed Missiles & Space, a visiting professor at the Software Engineering Institute of Carnegie Mellon University and at Florida A&M University, and a faculty member at East Tennessee State University. Professional memberships include ACM, IEEE, and the Consortium for Computing in Small Colleges.

Abraham Kandel, Ph.D., received his B.Sc. in Electrical Engineering from the Technion, Israel Institute of Technology, M.Sc. from the University of California, and Ph.D. in Electrical Engineering and Computer Science from the University of New Mexico. Dr. Kandel is a Professor and Endowed Eminent Scholar in Computer Science and Engineering and is Chairman of the Department of Computer Science and Engineering at the University of South Florida. Previously, he was Professor and Chairman of the Computer Science Department at Florida State University as well as the Director of the Institute of Expert Systems and Robotics at FSU and the Director of the State University System Center for Artificial Intelligence. He is a Fellow of the IEEE, a member of the ACM, and an Advisory Editor to the international journals *Fuzzy Sets and Systems, Information Sciences, Expert Systems,* and *Engineering Applications of Artificial Intelligence.* Dr. Kandel has published over 250 technical research papers and 19 books in computer science and engineering.

TABLE OF CONTENTS

1 INTRODUCTION

1   INTRODUCTION

I. QUALITY ASSURANCE FOR EXPERT SYSTEMS

Verification and validation are integral processes in software quality assurance. Simply stated, verification is the assurance of technical correctness; validation is the assurance of customer acceptance. Methods, techniques, and tools are needed for the comprehensive and systematic verification and validation of rule-based expert systems; however, current expert system technology has no such methods, techniques, or tools.

Software quality assurance is a factor in the production of expert systems. Verification and validation provide methods, techniques, and tools for the assessment and assurance of quality in conventional software (i.e., nonexpert system software). The concepts of verification and validation are applicable to expert systems. However, verification and validation, as currently used in expert system development, suffer from misinterpretation and misapplication. Verification and validation are complementary, but distinct, approaches to the assessment and assurance of quality software. However, in expert system development, the activities of verification and validation are intermingled and, therefore, indistinguishable.

Currently, verification and validation testing of expert systems is performed at the end of the development cycle in an informal, arbitrary fashion. This limited testing consists of the execution of a small set of test cases and the comparison of the results of the test cases to the answers of a human expert. Both the execution and the comparison are performed manually by the knowledge engineer (i.e., the designer and/or developer of the expert system) or the user. This analysis of the testing results is subjective since the results of the analysis may vary depending on the person performing the test case comparison.

The current approach to verification and validation of expert systems is partially a result of expert systems having existed primarily in the research environment. Verification and validation are unnecessary in a research environment where demonstrable proof of quality of the product is not required by a customer (instead, research aims towards proof of concept). With the enormous potential for expert system technology in commercial applications being recognized, expert systems are beginning to move from the research laboratory into industrial, government, and military use (i.e., into the commercial computing community). However, widespread acceptance of expert system technology has been slow. The commercial computing community, which is accustomed to the quality assurance methods of conventional software development, is reluctant to utilize this new technology without demonstrable assurance and assessment of its quality and reliability.

> What are all those people who have demanded (quite rightly) full acceptance testing of traditional software going to say when we approach them with a new type of computer system, which we cannot explain terribly well, and which comes up with answers the users do not understand? Especially when they find out that we have no evidence that anyone has ever fully explored the code or of what testing has been carried out on the system? [Oliver, 1987]

The move into commercialism requires, therefore, that expert systems be developed to standards equivalent to the established standards for software quality assurance. Verification and validation are approaches to satisfying these established standards and, thus, to advancing the generalized acceptance of expert systems.

II. BACKGROUND TO EXPERT SYSTEMS

Expert systems are "software systems (or subsystems) that simulate as closely as possible the output of a highly knowledgeable and experienced human functioning in a problem-solving mode within a specific problem domain." [Lane, 1986] The three main components of an expert system are the knowledge base (i.e., the expertise in a specific domain), inference engine (i.e., the controlling mechanism), and user interface (e.g., explanation facilities), as seen in Figure 1.1. The contents within the knowledge base are encoded by representational paradigms such as rules and frames. In a rule-based expert system, each rule is a unit of knowledge and has the basic form: 'if A and B then C' where 'A and B' is the antecedent clause and 'C' is the consequent clause.

Expert systems are a branch of artificial intelligence (AI). AI, which originated in the 1950s, is the field of computer science which endeavors to make "machines behave in a way that would generally be accepted as requiring human intelligence." [Prerau, 1990] By the 1960s, experimental work began on the concept of expert systems in research facilities at universities such as MIT, Carnegie-Mellon, and Stanford. Not until the 1980s did expert systems appear as commercial applications. Since the appearance of DECs XCON in 1981, the number of expert systems in commercial use has been increasing. Today there are approximately 3000 expert systems in use worldwide. [Fox, 1990]

However, 3000 is a relatively small number compared to the enormous amount of conventional software which is in use today. The slow acceptance of expert systems in commercial applications

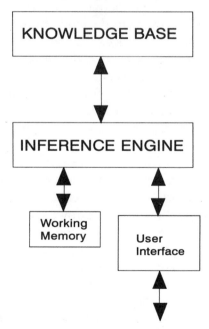

FIGURE 1.1. Components of Expert System

is due primarily to a lack of confidence in the reliability and quality of expert system software. Because the very nature of expert systems is unreliable since it is based on human knowledge or judgment and heuristic search techniques, the idea of producing software with such characteristics requires that "unless compelling evidence can be adduced that such software can be 'trusted' to perform its function, then it will not — and should not — be used in many circumstances where it would otherwise bring great benefit." [Rushby, 1988] Currently however, there exist no systematic methods, techniques, or tools in expert system technology for determining such "compelling evidence", i.e., determining the reliability or quality of expert system software.

Limited effort has been expended on developing approaches for assessing and assuring the quality of expert systems software since most expert systems existed only in the research laboratory environment until recently. The laboratory environment allowed expert system developers to be indifferent to issues facing commercial software developers such as project management, customer acceptance, maintenance, or legal liability. These laboratory expert systems were usually developed by a small, select team of highly skilled individuals. Furthermore, few of the expert systems in the research laboratory were developed beyond the prototype phase because "once the concept is demonstrated the work is finished." [Waterman, 1986] Finally, the approach to developing such expert systems, which has been called "programming by trial and error," leads to unreliable software. [Denning, 1986]

Several factors have led to a need for assurances about the quality of expert systems. One such factor is a new generation of expert systems which are beginning to appear. These expert systems are embedded within another hardware/software system in such a way that users may not even realize that an expert system is part of the software. Previously, expert systems were stand-alone applications built on special hardware with special software support. All input was received from the keyboard. These expert systems were used mostly as human assistants; they produced recommendations or advice which the human always monitored to ensure its accuracy or reliability before applying the recommendations or advice. Embedded expert systems, on the other hand, must receive input from hardware devices and/or other software applications, must operate within a real-time environment, and must be reliable enough to require no human intervention to oversee the accuracy of the recommendations or advice offered. For such expert systems to be widely accepted in the commercial computing community, expert system developers must be able to demonstrate a higher level of performance and reliability than was needed by previous stand-alone expert systems.

Another factor in the need for assurances about the quality of expert systems is that expert systems are being proposed for mission-critical applications. In such applications, there is a possibility of great financial loss and/or loss of life if the expert system malfunctions. Until the reliability and quality of such systems can be adequately exhibited, the commercial computing community will be reluctant to employ expert systems for mission-critical applications.

An additional factor is the expectations and standards for quality software which the commercial computing community has already established. These expectations and standards, such as customer-developer reviews and development documentation, are based on experiences from the development of conventional software. The commercial computing community, especially the Department of Defense, is unlikely to lower its expectations or standards of quality in order to adapt to expert systems. Expert system developers will have to adapt to the preestablished expectations and standards of the commercial computing community.

Time has arrived for expert system developers to address the issues concerning the production of quality software. Developers of conventional software faced this same problem in the late 1960s — a problem which they termed "software crisis" when the software being produced was unreliable, unmaintainable, and cost overrun. The developers of expert systems can benefit from the lessons learned as a result of the software crisis in order to avoid making these same mistakes.

III. CONVENTIONAL SOFTWARE AND EXPERT SYSTEMS

Merging the concepts of conventional software with the concepts of artificial intelligence and expert systems has been a lively topic of discussion in both fields of computer science. The differences between conventional software and expert systems have been widely reviewed. [Harmon, 1985, Ramamoorthy, 1987, Sarmiento, 1989] Only a few are described in Table 1.1.

In the early years of expert systems, more emphasis was placed on such differences. This differentiating attitude was especially strong while expert systems remained in the research environment. However, since expert systems have begun to move into the commercial computing community, the benefit of utilizing, where applicable, the concepts of conventional software has become more evident. Now the similarities between expert systems and conventional software are being explored in more detail in an effort to find concepts which are applicable to expert system development. In fact, some concepts of conventional software have already been utilized in the development of expert systems; these concepts include abstraction, information hiding, and modularity which have been applied to the design and construction of knowledge bases. [Harmon, 1990]

Table 1.1
Comparison of Expert Systems
and Conventional Software

Expert Systems	Conventional Software
Represent and manipulate knowledge	Represent and manipulate data
Symbolic information	Numeric information
Heuristic solution	Algorithmic solution
Requirements are vague	Requirements are well defined
Approximate answers	Exact answers
Control and data separate	Control and data intermingled

Expert systems and conventional software have many similarities. Both expert system developers and conventional software developers share the common goal of the production of quality software. This goal affects the development process, the quality assessment process, and other project planning processes. Additionally, much of the software within an expert system, such as the user interface and inference engine, is actually conventional software. Furthermore, both types of systems replicate a process, "whether that process is a thought process, in the case of most expert systems, or a procedure to perform some task or series of tasks" as with conventional software. [Sarmiento, 1989]

The question, then, is why merge the concepts of conventional software with those of expert systems. First of all, expert system developers are attempting to avoid "reinventing the wheel" on issues such as project management, software quality assurance, verification, and validation which are issues extensively addressed in the production of conventional software. Additionally, the developers of expert systems do not have the luxury of time for the evolution of such engineering concepts for expert system development. Today's software users are very sophisticated and less tolerant of software that is not of high quality, i.e., user-friendly, reliable, maintainable. Furthermore, expert system technology is being proposed for use in nuclear power plants, aircraft systems, financial advisement, and other applications where mistakes result in costly consequences. Finally, the interest in incorporating expert systems into commercial applications is immediate so the developers must meet this need, especially with the threat of technological and, thus, financial dominance by the Japanese in this area.

IV. A PREVIEW OF THE WORK

The contributions of this book promote software quality assurance for rule-based expert systems. These contributions are a systematic, comprehensive approach to the verification and validation of rule-based expert systems and a comprehensive set of techniques and tools for effective verification and validation of rule-based expert systems. These techniques and tools are demonstrated in the prototype SAVES (Suzanne and Abe's Validator for Expert Systems). SAVES consists of three main components which are the test manager, archiver, and analyzer and enhancer. The subsystem data flow diagram for SAVES containing these main components is shown in Figure 7.2.

The test manager of SAVES provides a user-friendly facility for the creation, manipulation, and execution of test cases. The results from the execution of test cases include the automated comparison of the expert system's results with the expected results. This result analysis is based on the statistical measures of order compatibility and distance comparability. The use of statistical measures eliminates the subjectivity of current verification and validation practices and the difficulty of evaluating certainty factors and multiple answers and adds formalism to the verification and validation of rule-based expert systems.

The archiver component within SAVES provides relevant information on previous testing performed by SAVES. This facility automates much of the tedious and time-consuming work of verification and validation. It also provides more systematic means of producing demonstrable proof of the accuracy and reliability of the expert system.

In addition to the execution analysis described above, the analyzer and enhancer facility within SAVES measures the effectiveness of the current suite of test cases in order to determine how completely this suite of test cases tested the knowledge base. Since the knowledge base is unique to expert systems, i.e., the inference engine and user interface are conventional software, a tool for measuring the effectiveness of the verification and validation of the knowledge base is beneficial. In addition to the analysis of this effectiveness, the analyzer and enhancer within SAVES provides suggestions on test cases. These suggestions are based on the test case generation techniques of input equivalence classes and output equivalence classes from conventional software development. This feature of SAVES complements the user-defined test cases, which are normally based on historical cases or real-world scenarios, with generated test cases in order to ensure more complete verification and validation of the knowledge base.

V. ORGANIZATION OF THE BOOK

The examination of verification and validation is organized in this book in the following manner. The concepts of conventional software are examined in Chapters 2 and 3. The considerable research and work on verification and validation for conventional software provide a pool of experience and understanding about the concepts of verification and validation and their capabilities and activities. Extensive examination of this work is central to understanding the potential for applying verification and validation to expert system technology. The whole picture of producing quality conventional software is addressed in Chapter 2. The concept and characteristics of quality software and approaches to producing quality software are discussed. The development process of conventional software is examined in Chapter 3. The functionality, inputs, and outputs of each phase of the classic life cycle process model are delineated and described in this chapter. Additionally, the verification and validation activities performed in each of these phases are examined. A detailed discussion of the verification and validation activities of conventional software development is included in Chapter 3.

The concepts of expert system development are examined in Chapters 4 and 5. The topics of these chapters parallel those of chapters 2 and 3. The concept and characteristics of quality expert systems are discussed in Chapter 4. The definition of verification and validation for expert systems, review of the current practices in expert system development emphasizing the verification and validation activities performed, and examination of recent research into verification and validation of expert systems are addressed in Chapter 5.

The weaknesses of the current practice in verification and validation of expert systems, lessons learned from previous experiences and difficulties in verifying and validating expert systems, and a systematic, comprehensive approach to the verification and validation of rule-based expert systems are discussed in Chapter 6. The implementation of a comprehensive set of tools and techniques which supports the approach, presented in this book, for the verification and validation of rule-based expert systems and the evaluation of this implementation are described in Chapter 7. Ideas for future research in the verification and validation of expert systems are presented in Chapter 8.

2

THE PRODUCTION OF
QUALITY SOFTWARE

2   THE PRODUCTION OF QUALITY SOFTWARE

I. INTRODUCTION

Concern about the quality of software is widespread and has resulted in much time, effort, and money being dedicated to quality assurance. There are several factors which make the production of quality software imperative. First and foremost, the use of computer software in mission critical areas such as nuclear control systems, space exploration, air-traffic control, and health care necessitates an exceptionally high level of quality.

As the use of computers has become widespread, there is a two-fold effect on the computer industry. One effect is that the computer industry has become a very lucrative and therefore, highly competitive business. In such a highly competitive marketplace, a reputation for quality is essential. The second effect is that since computers have become part of daily life, computer users are now more accustomed to computers and, consequently, are less tolerant of defective software. Additional pressure concerning quality is felt in the United States from the Japanese effort toward the production of quality software. Finally, producers of computers and software are being held legally liable for the quality of their products. [Zeide, 1990]

The meaning of quality when applied to software, the characteristics of quality software, and approaches to producing quality software are discussed in the following sections.

II. DEFINITION OF QUALITY SOFTWARE

The meaning of 'quality' is not readily apparent when applied to software. A typical dictionary definition of quality is "the degree of excellence of a thing." [Webster, 1987] Applying this general definition to software results in an ambiguous, incomplete definition of quality software. Historically, software projects were considered quality software if the errors in the code were unimportant. However, many times the software did not meet the specifications of the customer and, therefore, was useless; thus, the customer perceives the software to be of poor quality. Because of the frequency and costly consequences of such experiences, software quality has become more than just removing errors in the code. The many facets of quality software are conveyed in the following comprehensive definition:

1. The measure of acceptability of software products in relation to the application, environment, and role of the product in the context of the project.
2. The totality of features and characteristics of a software product that bear on its ability to satisfy given needs; for example, to conform to specifications.
3. The degree to which software possesses a desired combination of attributes.
4. The degree to which customers or users perceive that software meets their expectations. [Evans, 1987]

As seen in this definition, quality software is reflected in the software's fitness for purpose, fulfillment of requirements, and conformance to requirements. [Parrington, 1989] A more concise definition which also reflects the multiplicity of meanings for quality software is:

> Conformance to explicitly stated functional and performance requirements, explicitly documented development standards, and implicit characteristics that are expected of all professionally developed software. [Pressman, 1987]

The different perspectives of software which result in varying definitions for quality software are seen in the above definitions. These different perspectives are the consequence of the various people who are involved with the development, operation, and maintenance of software. Because "quality is in the eye of the beholder" [Cooper, 1979], the different points of view of these people provide several perspectives on the meaning of quality software.

The typical people associated with a software product during its development, operation, and maintenance are the user, i.e., the person regularly working with the software in its operational environment, the customer, i.e., the person ordering the software, the developer, i.e., the person designing and constructing the software product, the manager, i.e., the person monitoring the developer and the development process, and the maintainer, i.e., the person keeping the software in running order after it is in its operational environment. A user defines software quality as the user-friendliness, accuracy, and reliability of the software. A customer defines software quality as software that meets its specifications and is delivered on time and within budget. A manager defines software quality in terms of the manageability of the development process and the correctness and reliability of the product at the least cost. A developer defines software quality as the modularity, consistency, and testability of the software product. A maintainer defines software quality as the readability, understandability, modifiability, and expandability of the software product. The different perspectives of the people involved with a software product reveal that not only are the consistency, reliability, and correctness of the software product important but also the testability, maintainability, and manageability of the software development process.

III. ATTRIBUTES OF QUALITY SOFTWARE

As seen above, the definition of quality software is elusive. In fact, quality software is described throughout literature by a list of attributes commonly known as the "ilities" of quality software. [Boehm, 1978, McCabe, 1976, McCall, 1977] These attributes are well documented, although agreement on the exact list of attributes, and the definitions of these attributes may vary. Figure 2.1 shows the attributes defined by McCall. [McCall, 1977]

Reliability is the most prominent attribute of quality software. The IEEE definition [IEEE, 1983] describes reliability as "the ability

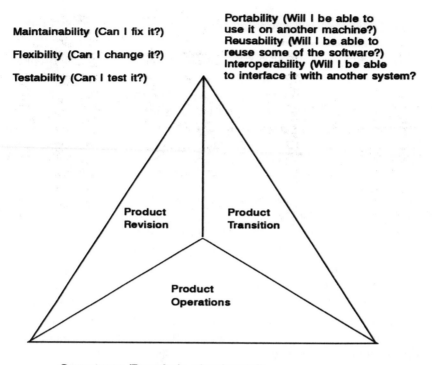

Maintainability (Can I fix it?)

Flexibility (Can I change it?)

Testability (Can I test it?)

Portability (Will I be able to
use it on another machine?)
Reusability (Will I be able to
reuse some of the software?)
Interoperability (Will I be able
to interface it with another system?

**Product
Revision**

**Product
Transition**

**Product
Operations**

Correctness (Does it do what I want?)

Reliability (Does it do it accurately all of the time?)

Efficiency (Will it run on my hardware as well as it can?)

Integrity (Is it secure?)

Usability (Can I run it?)

FIGURE 2.1. Software Quality Attributes

of a program to perform a required function under stated conditions for a stated period of time." The most common connotation of reliability is the robustness of the software, i.e., the software's ability to operate correctly regardless of the input or environment. However, reliability also refers to the software's ability to meet the user's requirements.

Other attributes which further detail the reliability of software include correctness, completeness, traceability, and consistency. Correctness refers not only to the extent to which the software is error free but also to the extent to which the software meets the specifications of the user, i.e., provides all the functionality desired by the user. Completeness pertains to the software having all the functionality as specified by the user and no extraneous functionality. Traceability is the ability to follow a requirement from the requirements specification through the design to code; traceability provides the means of assessing the completeness of the product. Consistency pertains to the uniformity within the products of the development process, e.g., the conformance to notation standards. Consistency and traceability are attributes which enhance the software development process.

Other attributes on the "ilities" list are testability — which encompasses the characteristics of understandability and measurability; maintainability — which encompasses modularity, readability, and simplicity; adaptability — which encompasses reusability, modifiability, expandability, and portability; and usability — which encompasses completeness, efficiency, integrity, and accuracy.

Often a customer does not know the attributes which are needed in his/her software system but can describe the characteristics needed for his/her software system. It is the responsibility of the professional software developer to recognize, and then implement, the attributes that satisfy such characteristics. For example, if the customer specifies that the software is to be used for many years, the developer realizes there will be many changes or corrections made to software which has a long life cycle; therefore, the software needs to have the attributes of maintainability, flexibility, and portability. If the software is to process classified information, then the software needs the attribute of integrity. If the software is a real-time application, then the software needs the attributes of efficiency, reliability, and correctness. [Walters, 1979]

Furthermore, as seen in an earlier definition of quality software, some of these attributes, e.g., measurability, expandability, are ones that the professional software developer realizes are part of quality software, even if the customer does not explicitly request these attributes. Thus, it is the responsibility of the professional software developer to determine which attributes are implicitly needed for a particular software project.

Finally, not all of these attributes for quality software are applicable to every software project. Some attributes are in conflict with other attributes; for example, increased portability or increased understandability may result in lowered efficiency. Additionally, constraints of schedule and cost may determine which attributes can feasibly be included in a particular software product. Ultimately, the customer and software developer must work together to define which attributes are essential to a particular project.

IV. APPROACHES TO ACHIEVING QUALITY SOFTWARE

Traditionally, software developers used testing as the primary approach to achieving quality software. However, as software systems became larger, more complex, and more critical in nature, testing proved to be inadequate as the only means of assuring the quality of software. One weakness to this approach was that testing was most often applied after the software product had been designed and coded which made errors discovered at that point very costly to correct. In the late 1960s, the computer community realized that to increase the quality of software required the establishment of methodologies which provide an effective software development process, the improvement of testing techniques, and the monitoring and review of the work and work products of the software development process. Some of the approaches which have been developed for these purposes are presented in the following sections.

A. Software Engineering Methodologies

Software engineering is a "systematic approach to the development, operation, maintenance, and retirement of software" with software being defined as "computer programs, procedures, rules and possibly associated documentation and data pertaining to the operation of a computer system." [IEEE, 1983] The concept of software engineering is a consequence of the software crisis of the late 1960s and the realization that there was a need for increased quality and productivity in the development of software. The problems encountered in the software crisis and the concepts of software engineering developed to eliminate these problems are illustrated in Figure 2.2. [Giarratano, 1989] Software engineering has transformed the development of software from the concept of programming to a disciplined approach which emphasizes early prevention or detection of errors, more visibility in the develop-

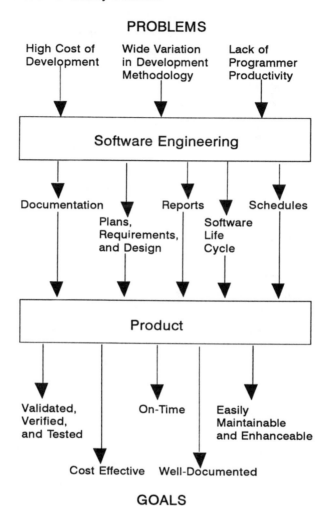

FIGURE 2.2. Software Engineering Methodology

ment process, and the assessment of quality and productivity in all stages of development.

The purpose of software engineering and the methodologies which support its concepts is to "engineer" quality into the software product by means of a systematic, incremental approach to development and the management of that development. In software engineering methodologies, systematic activities, supported by techniques and tools, coordinate the understanding and building of the software product. In addition, preliminary products are created as a result of these systematic activities. These products are used in assessment and management activities to assure that the requirements of the customer are being met, that specified standards are being followed, and, therefore, that quality is being engineered into the product.

The methodologies, which have been developed to support the concept of software engineering, include paradigms which provide the organizational techniques to development and management activities. These paradigms are the software engineering process models which present the phased approach to implementing organizational techniques. A software engineering process model is "a clear set of workproduct definitions and an indication of what steps should be taken to create those workproducts." [Freeman, 1987] Typical steps, i.e., phases, of a software engineering process model include analysis, design, implementation, and testing. Examples of process models are the waterfall model [Royce, 1970] and Spiral model. [Boehm, 1988]

Verification and validation, which are presented in detail in Chapter 3, are the major approaches to assessing quality in software engineering process models. Only a few of the assessment activities of verification and validation which occur in a typical process model are shown in Figure 2.3. [Fairley, 1985] As seen in Figure 2.3, the assessment of software is no longer performed exclusively at the end of its development, i.e., in the testing of the completed product, as done previously; instead, the activities encompassed in verification and validation are present throughout the phases of software engineering process models.

B. Software Quality Assurance

Software quality assurance is "a planned and systematic pattern of all actions necessary to provide adequate confidence that the item or product conforms to established technical requirements." [IEEE, 1983] Standards for quality assurance, i.e., "established technical requirements," were introduced into software development in the 1970s when government contracts specified such. Software quality assurance (SQA) was developed as the systematic approach to monitoring these standards.

The goal of SQA activities is to provide an objective analysis of the progress, process, and products of the development. To achieve this goal, SQA activities are present throughout the phased approach of the software engineering process models. These

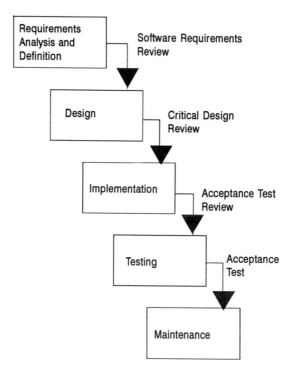

FIGURE 2.3.   V & V Activities

activities include monitoring compliance with standards, conducting formal technical reviews, testing the software, and reporting assessment information.

In order to provide greater objectivity in the reporting of assessment information, the SQA organization is managerially independent from the organization implementing the software engineering development process. The SQA organization reports its assessment information directly to the managerial level above the software project manager.

Although SQA activities do interact with the activities of verification and validation in the software engineering process model, the two groups implementing these activities are independent of each other. This independence adds extra assurance of the quality of the product by assessing that quality from two separate points of view. Verification and validation activities assess the technical aspects of the products (e.g., the operational behavior); SQA activities, on the other hand, assess the managerial aspects, e.g., quality of the end products of each phase. The relationships between SQA, the software engineering development process, and independent verification and validation are illustrated in Figure 2.4. [Pressman, 1987] As seen in Figure 2.4, the SQA group acts as a representative for the customer in the development process. One of the objectives of SQA activities is to ensure that the wants and needs of the customer are implemented in the software product.

C. Independent Verification and Validation

Independent verification and validation (IV & V) is utilized when the nature of the application is so critical that extra assurance of functionality and performance is required. IV & V activities include analysis and testing in order to assess the software's quality and to ascertain if the critical, i.e., high risk, requirements are achieved in the software. Although the IV & V activities are similar to those of verification and validation performed in the development process, IV & V is managerially independent from the organization(s) implementing verification, validation, and the development process. The cost of IV & V is high; often it is 25 to 30% of the development cost. The high cost is balanced by the criticality of the software being developed and the disastrous effect of that software being unreliable or erroneous.

D. Software Reviews

Evaluations of the products of the software development process are conducted in both software quality assurance activities

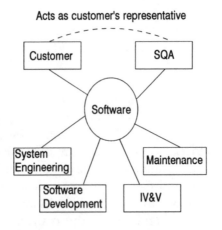

FIGURE 2.4. Software Quality Assurance

and verification and validation activities throughout the phases of the process model. These evaluations establish early assessment of the quality of the software product by "evaluating the process being used to develop the software, conformance of the software project to the needs of the program, and the technical requirements, goals, and objectives of the system." [Evans, 1987] These evaluations are achieved through formal reviews and internal reviews. Formal reviews occur at the end of each phase of the software life cycle, are performed on the end products of these phases, and result in formally approved products. Examples of formal reviews include system requirements review, software design review, and test review. Internal reviews provide, to members of the development organization, a means of examining and analyzing draft versions of a software product or a portion of a software product. Three techniques for software reviews are walkthroughs, inspections, and audits.

Walkthroughs are group meetings where a product is examined by a team of reviewers. The engineer of the product leads the team of reviewers through a detailed examination of the product. Feedback is provided by the team of reviewers on any problems or defects found in the product and on any improvements in structure or style. The suggestions on problems or improvements are itemized, but their solution is delegated to the engineer of the product.

An inspection is a more formal, rigorous evaluation and reporting technique. Someone other than the engineer of the product supervises the team of reviewers performing the inspection. Each member on the team of reviewers has a particular role in the evaluation process and is working from a checklist of errors. If problems or defects are found in the product, a solution is proposed by the team.

An audit is a non-team activity to evaluate the compliance of the process or product with established standards or requirements. Audits are usually conducted by sources outside of the development process.

E. Formal Proof of Correctness

Formal proof of correctness, which is also known as formal verification, is "a rigorous mathematical demonstration that source code conforms to its specifications." [IEEE, 1983] Three methods of formal proof of correctness are input-output assertions, weakest precondition, and structural induction. Some in the computer community believe that this technique is the only approach that can completely guarantee (via the rigor of mathematics) consistency between the software solution and the requirements specification. However, formal proof of correctness has not been widely accepted in the computer industry because it is extremely difficult and costly to use.

Proof of correctness has been utilized on limited segments of critical software where the high risk of loss of human life balances the great expense of applying proof of correctness techniques. Interest in formal verification is being generated by the work on the Cleanroom process [Mills, 1987] which emphasizes the use of formal verification with software engineering. Research into the automation of formal proofs of correctness is also ongoing.

F. Quality Metrics

The objective of quality metrics is to provide quantifiable measures of the quality of a software product. Such metrics facilitate the software engineering development process by providing quantifiable assessment to guide the development. However, defining quantifiable measures for the attributes associated with quality software, discussed in Section III of this chapter, has proven to be very difficult because of the subjective nature of these attributes. Attempts [Boehm, 1978, McCabe, 1976, McCall, 1977] have been made to describe these attributes in a quantifiable manner. For example, the measurement of the cyclomatic complexity of a software component [McCabe, 1976] attempts to quantify the maintainability of that component. Research into quality metrics and their role in the production of quality software is ongoing.

G. Configuration Management

Configuration management plays an active role in the production of quality software. Configuration management activities parallel the development and maintenance of software. During development and maintenance of software products, changes are frequently introduced in these products. An organized approach to monitoring these changes is required to maintain quality software. A primary objective of configuration management is to provide this control of change. After a software product is approved in a formal review, that product is then placed under the control of configuration management. Any future changes to that product are supervised; for example, a change is performed only if approval is given. A product under the control of configuration management is called a baseline. Configuration management also maintains and reports the status of all software products which are baselined and monitors the release of new versions of software products.

V. APPLICATION OF APPROACHES

In order to produce quality software, the approaches of software engineering methodologies, software quality assurance, IV & V, and others described in Section IV of this chapter must be tailored into a coordinated, systematic methodology which satisfies the needs of a particular software project. No one approach is effective if used in isolation. The concepts and practices supported by software engineering are essential to the production of quality software because it is the only approach to address quality during the development of the software. An example of a software engineering paradigm, the waterfall model or classic life cycle process model, supplemented with quality software approaches of verification and validation is presented in Chapter 3.

3

CONVENTIONAL SOFTWARE DEVELOPMENT

3

CONVENTIONAL
SOFTWARE DEVELOPMENT

I. INTRODUCTION

The development of conventional software is supported by the systematic, disciplined approaches of software engineering. A process model is a software engineering paradigm which represents the activities of the software life cycle, i.e., the development and maintenance of a software product. One of the best known and perhaps best understood process models is the classic life cycle or waterfall model. This process model represents the software life cycle as a series of distinct, sequential activities where each activity has specific input(s) and specific output(s) and where the output of one activity is the input to the next activity in the sequence. These specific inputs and outputs, i.e., the documents of the classic life cycle process model, act as catalysts for the activities of this process model; consequently, the classic life cycle process model is referred to as a document-driven process model.

Often considered too simplistic for practical software development, the classic life cycle process model, nevertheless, satisfactorily depicts the technical and managerial activities commonly utilized in the software life cycle for the production of quality software. The following sections describe the inputs and outputs, i.e., documents, the functionality, and the verification and validation activities for each phase of the software life cycle.

II. VERIFICATION AND VALIDATION

Verification and validation are major approaches within software engineering process models for ensuring the production of quality software. In the past before verification and validation, software developers used testing as the primary approach to achieving quality software. The main goal of this testing was the removal of errors in the code. Such testing was usually performed in an informal, arbitrary manner by programmers who worked, for the most part, in isolation from the rest of the development team. Furthermore, the testing was done mainly at the end of the development process where errors are very costly to correct. As software systems became larger, more complex, and more critical in nature, this type of testing proved to be inadequate. People realized that quality could not be tested into a software product at the end of the development process and thus the issue of quality needed to be addressed throughout the development process. [Fairley, 1985] As a result, comprehensive approaches to ensuring quality software, called verification and validation, were developed.

Verification and validation are present throughout all the phases of the software life cycle and are composed of a wide variety of activities. The diverse activities of verification and validation activities allow for the early detection and correction of errors. In addition, the incremental assessment, which verification and validation activities provide at each phase of the process model, has become a managerial benefit. The information resultant of these activities gives a progressive view of the productivity of the development process and the quality of the software product. Furthermore, many techniques and tools have been developed in recent years which support the activities of verification and validation. These techniques and tools enhance the effort toward achieving quality software through formalism and automation.

The activities of verification and validation are often so closely tied that verification and validation are commonly known as one task called V&V. Collectively, V&V has the technical goals of assessing the quality of the products developed in the software life cycle and determining whether these products satisfy the specifications of the customer. However, each has its own distinct objective towards achieving these goals and its own set of activities for accomplishing that objective.

Verification is "the process of determining whether or not the products of a given phase of the software development cycle fulfill the requirements established during the previous phase." [IEEE 1983] Accordingly, verification activities evaluate the end product of each phase of the process model not only to determine its consistency, completeness, and correctness but also to determine its compliance with the actions and products of the preceding phase of the process model. Verification, in this way, attempts to reduce or eliminate errors generated within a phase and by the transformation of information between two phases. In other words, verification ensures that the development activities within a particular phase of the software life cycle are correctly implemented. Succinctly stated, verification answers the question "Am I building the product right?" [Boehm, 1984b]

Validation is concerned with demonstrating whether or not the software product actually solves the customer's problem. Validation is "the process of evaluating software at the end of the software development process to ensure compliance with software requirements" [IEEE, 1983]. Validation assesses the end product to determine its compliance with the software and system requirements specified by the customer at the beginning of the development process. Validation answers the question "Am I building the right product?" [Boehm, 1984b]

Verification and validation achieve their quality objectives with a combination of three types of analysis. Static analysis, dynamic analysis, and formal analysis are utilized by verification and validation to obtain three different approaches to the assessment of the quality of software products. Static analysis, which does not involve the execution of software, is the manual or automated examination of a software product, i.e., requirements, design, code. For example, the different products of the software life cycle phases are compared to determine the consistency between these products. Walkthroughs and inspections are examples of techniques used in static analysis. Dynamic analysis is the execution of software to evaluate the functional, structural, or computational aspects of the software. Unit testing, design unit testing and system testing are examples of dynamic analysis. Formal analysis is the use of mathematical techniques to evaluate a software product, e.g., analysis of an algorithm. Symbolic execution and proof of correctness are techniques of formal analysis. These three forms of analysis provide in the verification and validation activities a comprehensive evaluation of the software's quality. The activities of verification and validation are comprehensively discussed in the following sections along with the inputs, outputs, and functionality of the phases of the classic life cycle process model.

III. OVERVIEW OF CLASSIC LIFE CYCLE PROCESS MODEL

Several versions of the classic life cycle process model are used in industry today. All these versions encompass the definition, development, and maintenance of the software product. The traditional phases of the classic life cycle are the requirements analysis and definition phase, design phase, implementation phase, testing phase, and maintenance phase. Prior to these phases of the software life cycle, the analysis of the system in which the software is to exist is performed. During this phase, which is often referred to as system requirements analysis, the whole environment of the system including the hardware, people, and other software is examined. The requirements of each of these elements within the whole system are gathered and analyzed in order to better understand the environment in which the software is to function. Once the overall system is understood, then the software life cycle begins. The software life cycle including the system requirements analysis phase are depicted in Figure 3.1.

A. Requirements Analysis and Definition Phase

The essential activity within the requirements analysis and definition phase is the collection and understanding of the customer's requirements, i.e., the customer being the person(s) contracting the software development and/or the person(s) working regularly with the software in its operational environment. Greater emphasis has been placed on this activity, particularly in the customer's involvement in this activity, for it has been estimated that 45% of the problems discovered previously in the testing phase were the result of incorrect or incomplete software requirements. [Cooper, 1979] This activity requires extensive communication between the

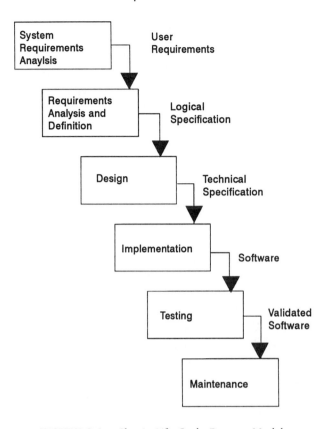

FIGURE 3.1. Classic Life Cycle Process Model

customer and the developer. These requirements include the functionality of the software, any constraints with which the developer must deal, and the performance expected of the software. The outcome of this work is the development of a requirements specification document. Figure 3.2 is an outline of the information typically provided in this document. The completeness and accuracy of this document is very important since it serves as the basis for the remaining phases of the software life cycle. Furthermore, this document often acts as a contract between the customer and the software developer by specifying what the developer has agreed to do for the customer. Liability, as stated earlier, is such a growing concern in software development that the legal validity of the requirements specification document is now an issue. "Even though there may not be the slightest chance of the specification forming the basis for a lawsuit, ... specifications should always be written as if they will be used as evidence in a

1 Introduction
1.1 System reference
1.2 Business objectives
1.3 Software project constraints

2 Software description
2.1 Objectives and operations
2.2 Flow model
2.3 Data dictionary
2.4 System interface description

3 Processing narratives
3.n Transform n description
3.n.1 Processing narrative
3.n.2 Restriction/limitations
3.n.3 Performance requirements
3.n.4 Design constraints
3.n.5 Supporting diagrams

4 Validation criteria
4.1 Testing strategy
4.2 Classes of tests
4.3 Expected software response
4.4 Special considerations

5 Bibliography
6 Appendix

FIGURE 3.2.   Framework for Software Requirements Specifications

trial." [Schach, 1990] Other activities within this phase include conducting a feasibility study, developing a project plan document which specifies the management strategy for the development process, and creating a preliminary user manual.

Unlike software development of the past, verification and validation activities begin in the requirements analysis and definition phase of the classic life cycle process model. The primary verification task is the formal review of the requirements specification document. Participants in this review include the developer and customer. Customer representation in this review is often comprised of several types of software users. The three types of users commonly involved in this formal review are the primary user who is the member of the customer organization regularly using the software, the secondary user who is part of the customer organization and shares access to information generated or maintained by the software, and the tertiary user who is outside of the customer organization but periodically requires access to the information generated or maintained by the software. [Parrington,

1989] The participation by the customer in this formal review is extremely important. As stated earlier, often the requirements specification document acts as a legal contract between the customer and the software developer; therefore, no misunderstandings or ambiguity is acceptable in this document.

In the formal review of the requirements specification document, the document is examined for its completeness, consistency, testability, correctness, and feasibility. The requirements specification document is complete if everything is present and fully specified, e.g., there are no "To Be Determined" items, no missing functions. The requirements specification document is consistent if nothing in it conflicts with the system requirements and no part within it conflicts with another part of the requirements specification document. The requirements specification document is testable if the specifications are well-defined and measurable so that a pass or fail assessment of the implementation of the specifications can be made during validation. The requirements specification document is correct if the specifications accurately reflect the system requirements and the customer's expectations. The requirements specification document is feasible if the "life-cycle benefits of the system specified exceed its life-cycle costs." [Boehm, 1984a]

Within this requirements specification document is a section essential to the validation of the software product. The acceptance criteria section defines the characteristics of the software system which are to be demonstrated in order to declare the system successful. In validation testing performed during the testing phase of the classic life cycle, the developer must show that these acceptance criteria have been successfully satisfied before the software is acceptable to the customer. Thus, specifying these criteria in unambiguous, measurable terms is integral to the validation process.

Other verification and validation activities within the requirements analysis and definition phase include the development of a Project V & V Plan, generation of requirements-based test cases, review of the preliminary user manual, and review of the project plan document. The Project V & V Plan outlines the verification and validation activities applicable to this software project and delineates the techniques and tools to be used and the schedule to be followed in order to perform these verification and validation activities. The test cases, generated from the requirements, form the basis for validation testing performed later during the testing phase of the classic life cycle. These test cases are designed to demonstrate that the right product is being built. This repository of test cases is augmented during each succeeding phase of the process model.

B. Design Phase

The requirements specification document generated within the requirements analysis and definition phase acts as the catalyst for the design phase. During the design phase, the logical representation of the system, as described in the requirements specification document, is systematically transformed into the physical representation of the system. In other words, the requirements tell the developer "what to do" while the design tells the developer "how to do it."

The design phase is normally a two-part process. The first part is the preliminary design. In this part of the design phase, the developer prepares a high-level view of the architectural structure of the software solution. This architectural structure is a hierarchical representation of the modules which satisfy all the functionality specified in the requirements specification document. The second part of the design phase is the detailed design. Here the intermodule communication, data structures, and algorithmic details are specified for each module of the preliminary design. The goal of these two parts is to create an algorithmic solution that can be directly translated into a programming language in the implementation phase and that clearly achieves the functionality delineated in the requirements specification document.

Other activities of the design phase are the revision of the requirements specification document and the project plan document if discovery of problems or errors during the design phase warrants such revision. Any change to these documents requires approval and additional review since the documents are baselined, i.e., turned over to configuration management, at the end of the requirements analysis and definition phase. Additional activities include the adoption of standards, e.g., notational standards for the design or code, and the design of databases necessary for the software.

The primary verification activity during the design phase is the formal review of the preliminary and detailed design documents.

These documents are verified for their consistency, completeness, and correctness within themselves and with the requirements specification document. The software design is evaluated to ensure that all the functionality of the requirements specification document is present and that no extraneous functionality is present. The customer is usually not present at this review.

Other verification and validation activities during the design phase include the generation of design-based test cases and the completion and review of the Project V & V Plan. Also specified during the design phase is a test plan for the modules which are delineated in the design. This test plan includes guidelines for testing of the individual modules, guidelines for the integration testing of these modules, a schedule of these testing activities, and the resources and personnel needed to perform these tests. This test plan is reviewed along with the design documents.

C. Implementation Phase

The implementation phase is the translation of the detailed design into source code. Ideally, if the analysis and design of the solution are well done, this translation is a very mechanical activity.

The verification and validation activities of the implementation phase include the review and testing of the code of the individual modules. The code is reviewed to determine its completeness, correctness, and consistency with the detailed design. Additionally, the software units within the code are traced back to the requirements specification document to ensure that each software unit is necessary code, to ensure that no extraneous functionality has been introduced into the code, and to ensure that all the requirements delineated in the requirements specification document are present in the software. The code is also evaluated for its conformity to specified standards for both internal documentation and programming style. The repository of test cases which has been built in the requirements analysis and definition phase and the design phase is completed during the implementation phase.

D. Testing Phase

The activities of the testing phase revolve around the controlled execution of part or all of the software product. Testing is the best understood part of the verification and validation process.

The verification and validation activities of the testing phase are performed to demonstrate that the functionality of the requirements specification document has been met, to detect any remaining errors in the software, and to evaluate the reliability of the software.

A goal of verification and validation in software engineering methodologies is to eliminate as many errors as possible before the testing phase of the software life cycle. To correct errors such as misunderstood, missing, or incorrect requirements during the testing phase costs 100 times more than if these errors were corrected in the initial phases of the software life cycle. [Boehm, 1984a] However, even with the systematic approach to verification and validation performed in the requirements analysis and definition phase, the design phase, and the implementation phase, errors may still exist in the software product. Therefore, a well defined, structured approach to exercising the software is established in the test documents, i.e., the Project V & V Plan and the test plan. Providing this structured approach to testing furnishes extra confidence in the quality of the software.

The four major types of testing activities which comprise this structured approach to exercising software are unit testing, design

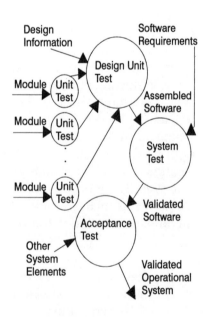

FIGURE 3.3.   Testing Activities

unit testing, system testing, and acceptance testing. The first three activities are part of the verification process for the software. The last two activities combine to perform the validation of the software. System testing is utilized in both verification and validation. The relationship between these four testing activities is illustrated in Figure 3.3. [adapted from Pressman, 1987]

Unit testing which is partially, if not completely, performed during the implementation phase is the exercising of individual units of code. A unit of code is tested in isolation from the other parts of the software. This testing is usually performed by the person who developed the unit of code. Unit testing exercises the internal structure of the code, i.e., exercising each statement, the conditional statements, the loop conditions. Unit testing is also termed as module testing.

In design unit testing, units of code are assembled according to the design documents, and then the assemblages are tested as specified in the test plan. Each of the units of code in the assemblage has already been tested individually in a unit test. Thus, design unit testing is concerned with the interfacing between the units of code which have been grouped, with the functionality performed by the units of code as a group, and with the effects on global data, e.g., files, databases. Also known as string testing or integration testing, design unit testing integrates the units of code into a group using either a top-down strategy, bottom-up strategy, or a combination of the two, called sandwich strategy.

System testing begins when the entire software product has been integrated and exercised through design unit testing. System testing is performed by the development group in the development environment, instead of the software's operational environment. System testing performs both verification and validation. As a part of verification, system testing is used to eliminate errors in the entire software product. As a validation activity, system testing allows the developer to evaluate the entire software product against the functional and performance requirements which are enumerated in the requirements specification document.

Acceptance testing is performed in the operational environment of the software and is often performed by someone outside of the development group such as the customer, a SQA group, or an IV & V group. The goal of acceptance testing is to demonstrate that the software satisfies all the acceptance criteria specified in the requirements specification document. This testing is the last before the software is handed over to the customer which is the beginning of the maintenance phase of the software life cycle.

E. Maintenance Phase

Once the software system is turned over to the customer, the activities of the maintenance phase begin. The activities in this phase include the processing of software modifications and the continual evaluation of the software. During the maintenance phase, modifications arise due to enhancements requested by the customer, change in the operational environment, and discovery of errors. All modifications to the software product are strictly monitored and require adherence to procedural standards for implementing changes.

The primary verification and validation activity during this phase is regression testing. If any modification is made to the software, regression testing is performed to revalidate the software. Regression testing serves to evaluate a modification's effect on the rest of the software and on the acceptance criteria of the requirements specification document. The test cases developed during the requirements analysis and definition phase, design phase, and implementation phase are often reused and augmented during regression testing.

IV. RESULTS OF USAGE

The use of verification, validation, and software engineering concepts in the development of software have resulted in numerous benefits. A primary benefit is the early detection of errors. Errors such as missing or misunderstood requirements are very costly to correct if not detected early in the software life cycle. Another benefit is that verification and validation require explicit development planning throughout the life cycle which results in better understanding of the requirements, better control of the development schedule, and better control of the testing activities. The visibility into the quality of the product and the development process which the activities of verification and validation furnish is a benefit to project management. An additional benefit is that the requirements of the customer are now considered a primary factor in the development process and in verification and validation. This focus on requirements has resulted not only in a better understanding of the requirements but also in more complete and accurate requirements.

Results from utilizing verification and validation in the development of software are not widely documented. In one recent report [Wallace, 1989], two studies presented different views on the effectiveness of verification and validation. The Software Engineering Laboratory at the National Aeronautics and Space Administration's Goddard Space Flight Center reported little effectiveness in three projects where verification and validation were employed. The three flight-dynamics projects were medium-sized projects of 10,000 to 50,000 lines of source code with a development life of 18 months and with a development team of one to three people per project. The primary positive result of utilizing verification and validation on these projects was a better rate of error detection early in the development process.

The Air Force's Rome, New York Air Development Center presented much more favorable results from the utilization of verification and validation. The four projects, which were real-time command-and-control, missile-tracking, and avionic programs, were large-sized projects of 90,000 to 176,000 lines of source code. The development life on these projects was 2.5 to 4 years with development teams of 5 to 12 people per project. Much improvement was seen in the early detection of errors and in programmer productivity. For the Air Force projects which utilized verification and validation throughout the life cycle, the savings from early error detection were 92 to 180% of the costs of using independent verification and validation. A preliminary conclusion that may be drawn from these two examples is that the effectiveness of verification and validation is better seen on larger projects which deal with a longer development life and a larger development team.

Another recent report on the effectiveness of software engineering [Taylor, 1989] states that results from several development projects including a nuclear power plant control system show that reliability was improved by a factor of 10 with the use of software engineering concepts. This report also relates that the majority of producers of safety-related applications employ software engineering.

V. SUMMARY OF VALIDATION IN CLASSIC LIFE CYCLE

Although validation is defined as "the process of evaluating software at the end of the software development process to ensure compliance with software requirements" [IEEE, 1983], the activities of validation occur throughout the software life cycle, not just at the very end as the definition implies. Each of these activities is essential to achieving the complete validation of a software product.

A. Validation Testing

Validation testing demonstrates that the software product complies with the stated requirements of the customer. Dynamic analysis is utilized to perform this demonstration. The types of dynamic analysis used are system testing and acceptance testing. System testing, which is performed in the development environment, demonstrates to the development team that the software product complies with the requirements of the customer. System testing, which is performed by members of the development team, is investigatory in its effect to expose problems or errors in the software product. System testing, therefore, has two functions which makes it a verification activity as well as a validation one. These functions are:

1. to validate that the completed software functions close enough to its requirements to warrant passing it on to the next stage in the life cycle and
2. to improve the quality of the completed software by first discovering the input states on which it fails to meet requirements and then eliminating these failures by correcting the underlying faults. [Musa, 1989]

Acceptance testing, which is performed in the operational environment, demonstrates to the customer that the software product complies with his/her requirements. Acceptance testing, which is usually performed by someone outside of the development team, e.g., customer, IV & V, SQA, is more demonstrative in its effort to show that the software functions properly. [Ould, 1986] The goal of system testing is to exercise the software product so thoroughly that no surprises or problems arise during acceptance testing; in other words, acceptance testing is always expected to be successful if system testing is done well. In fact, acceptance testing should be just formal reassurance to the customer that the software performs what and how he/she expected. [Hetzel, 1988]

B. Validation Test Cases

Validation testing, i.e., system testing and acceptance testing, is a technique of dynamic analysis. Dynamic analysis, which is the execution of software to evaluate functional, structural, or computational aspects of the software, requires the preparation of test cases and expected results. Test cases are developed throughout the software life cycle in the requirements analysis and definition phase, the design phase, and the implementation phase. Test cases developed during the requirements analysis and definition phase are designed to demonstrate the functional and performance requirements and any constraints described in the requirements specification document. The test cases developed in the design phase are designed to exercise the elements of the design such as the algorithms. The test cases developed in the implementation phase are designed to exercise the structure of the source code. [Powell, 1986]

One of the major concerns with the generation of test cases is the balance between having enough test cases to thoroughly exercise the software and having more test cases than can be adequately processed. Exhaustive testing, which is the exercising of a software system on all possible inputs, is economically and computationally impossible for most software. Therefore, the objective is to develop a set of test cases which is comprehensive enough to adequately demonstrate the correctness of the software and which is small enough to be adequately generated, executed, and analyzed.

Functional analysis and structural analysis are applied to the software products in order to develop such a set of test cases. Functional analysis, which is also known as black box testing, data-driven testing, or functional testing, is used to design test cases which focus on the requirements. The emphasis of these test cases is the demonstration that all functionality as specified in the requirements specification document is exercised, that all categories of valid input are accepted, and that all categories of invalid input are rejected. Techniques for deriving such test cases include equivalence partitioning, boundary value analysis, cause-effect graphing, and error guessing.

Structural analysis, which is also known as white-box testing, glass-box testing, metric-based testing or logic-driven testing, is used to design test cases which focus on the internal structure of the software. For example, a type of structural analysis is branch testing where test cases are generated to exercise each branch in a segment of software. The emphasis of structural analysis test cases is to exercise the software to a specified level of thoroughness which is usually measured by some metric. The most commonly used metrics for structural analysis are statement coverage, branch coverage, and path coverage. To aid in measuring these types of coverages, an automated tool called test coverage analyzer is utilized to record data about the code which is exercised by a set of test cases. The test coverage analyzer provides information such as the percentage of coverage and segments of code which are not exercised by the set of test cases. The information supplied by this tool reveals the effectiveness of the set of test cases.

Structural analysis complements functional analysis by exercising segments of software which were not exercised by the test cases generated by functional analysis. System testing uses both functional analysis and structural analysis techniques for the production of test cases while functional analysis produces the test cases which are used in acceptance testing.

C. Acceptance Criteria

". . . in order to achieve high-quality software, we must specify what we mean by high quality and then implement that specification. In other words, High quality = Meeting the quality specification." [Deutsch, 1988] Each customer has a different view of quality software. Therefore, having the customer specify exactly what is needed, wanted, and required to satisfy his/her view of quality software for a particular project is a necessity if validation is to demonstrate that the final software product complies with what the customer expected. The determination and establishment of these requirements are accomplished in the requirements analysis and definition phase of the software life cycle and are evidenced in the requirements specification document. This document is reviewed and agreed upon by both the customer and the developer in the formal review at the end of the requirements analysis and definition phase and is then baselined. The emphasis on the generation and approval of this document is a result of software systems which have failed due to misunderstandings between the customer and the developer. Furthermore, the rest of the development process is based upon this document. The objective in generating this document is to eliminate ambiguity by providing well defined, testable requirements which are demonstrable by validation testing.

Part of this requirements specification document, as seen in Figure 3.2, is the acceptance criteria upon which validation is based. These criteria are the requirements by which the customer judges the software to be acceptable. These criteria include functional requirements, performance requirements, standardization requirements for any documentation products, requirements for hardware, software, and user interfaces, requirements for quality assurance standards, and requirements for the development schedule or process. Without these acceptance criteria, validation is just a guessing game. In other words, good validation requires a good requirements specification document.

D. Testing Documentation

Two types of testing documentation are produced during the software development process. These two are the Project V & V Plan and test plans. The Project V & V Plan is the comprehensive plan of action for the verification and validation activities of a particular software project. As discussed previously, the Project V & V Plan is generated in the requirements analysis and definition phase and finalized in the design phase. The Project V & V Plan "names the members of the engineering staff who have specific V & V responsibilities, defines QA responsibilities, provides a schedule of V & V activities keyed to events identified in the development plan, defines the intended V & V tasks and reports, and describes the procedures that will be followed." [Dunn, 1990] In other words, it specifies all the project-specific activities, responsibilities, and scheduling of verification and validation. Additionally, the Project V & V Plan identifies techniques and tools to be used to perform the specified verification and validation activities.

Test plans are generated during the design phase of the software process model and finalized in the implementation phase.

Test plans are produced for each of the major testing activities which are unit testing, design unit testing, system testing, and acceptance testing. A test plan is "an overall plan and schedule indicating which specific tests are to be conducted and how test results are to be documented." [Eliason, 1990] Issues which are addressed in a test plan include the testing objectives, e.g., items or features to be tested, test completion criteria, e.g., percentage of coverage, milestones and schedule, testing personnel and their responsibilities, description of test cases to be performed, resources needed to perform these test cases, and description of the test analysis documentation which presents the test results. The test plan also includes the test scenarios or scripts and the expected results in the description of the test cases.

Of particular importance to validation is the Acceptance Test Plan. The Acceptance Test Plan is created by the development team and the customer. It describes the tests which, if successfully executed, demonstrate that the software product is acceptable to the customer. An example of the format of an Acceptance Test Plan is:

Section 1: Requirements to be verified
Section 2: Test cases for each requirement
Section 3: Expected outcome of each test case
Section 4: Capabilities demonstrated by each test [Fairley, 1985]

All testing documents developed within the software life cycle are reviewed and baselined. These reviews are to ensure that the contents of the testing documents are feasible and consistent with the proposed development plan.

E. Revalidation During Maintenance Phase

The revalidation of a software product is necessary when any change or modification to the software is implemented during the maintenance phase. Revalidation, which is also termed regression testing, demonstrates that the software product still satisfies the customer's requirements. This type of testing utilizes the functional analysis test cases which were generated in the requirements analysis and definition phase and have been maintained in project documentation. New test cases which are needed to exercise the change or modification are also added to this repository of test cases. Utilizing the existing repository of test cases for revalidation testing saves the time, money and effort required to generate a new suite of test cases to effectively validate the software.

F. Validation Tools

In recent years many semi-automated and automated tools have been developed to facilitate the verification and validation process. A comprehensive list of the tools which are utilized in the verification and validation process [Wallace, 1989] is presented in Appendix A. Some of the tools which are specific to validation include test coverage analyzers, comparators which are used to compare a file containing expected output with the generated output, test data generators, simulators which duplicate the operational environment of software being tested, and testing information systems which retain test cases and testing status. [Hetzel, 1988] Many of these tools assist in the management of the validation process, especially in the bookkeeping tasks related to validation testing, e.g., maintaining the test cases and the test results, analyzing history of testing.

4

THE PRODUCTION OF
QUALITY EXPERT SYSTEMS

4

THE PRODUCTION OF
QUALITY EXPERT SYSTEMS

I. INTRODUCTION

The production of quality software is a goal shared by both conventional software developers and expert system developers. Factors such as the criticality of applications, the competitive and lucrative marketplace, the expectations of sophisticated users, and legal liability, as discussed in Section I of Chapter 2, have created the concern about the quality of software. At the present time, these pressures may not be as evident for expert systems as for conventional software since the number of companies developing conventional software and the marketplace for conventional software are more extensive than for expert systems. However, expert system developers have even more reason for striving to produce quality software, for they are attempting to emulate a human expert. The expectations for human experts are great, but the expectations for software systems imitating human experts are often even greater. People are normally tolerant of mistakes in human experts; after all, they are just human. However, tolerance of mistakes in computer software is much lower.

Expert system developers are realizing that in order to compete in the commercial computing community an expert system must display quality not only in its performance but also in its structure and development process. Furthermore, expert system developers are recognizing that the quality of performance must be demonstrated in a systematic, well documented manner in order to promote customer acceptance of this new technology in commercial applications. The following sections examine the meaning of quality unique to expert systems, the characteristics of quality expert system software, and approaches to producing quality expert system software.

II. DEFINITION OF QUALITY EXPERT SYSTEMS

As seen in Section II of Chapter 2, the concept of quality software has been widely examined and analyzed for conventional software. Most of the definitions discussed for conventional software apply directly to expert systems since expert systems are software and much of an expert system, i.e., user interface, inference engine, is conventional software. However, many aspects of these definitions of quality software have never been addressed by expert system developers. Documentation standards, compatibility within an operating environment, and conformance to implicit standards of professional software developers are examples of issues rarely associated with expert systems. Not only these issues but also the issues related to the unique aspects of expert systems need to be considered when defining quality expert system software.

A unique aspect of expert systems is the emulation of human experts. Thus, a definition of quality for expert systems is the ability of the expert system "to equal the performance of human experts." [Hu, 1987] Another unique feature of expert systems is the separation of the knowledge and control where knowledge resides in the knowledge base and control in the inference engine. Since the inference engine is algorithmic software, i.e., conventional software, the definitions of quality for conventional software apply to it. The quality of the knowledge base needs to be considered. In fact, the quality of an expert system is often equated with the quality of the knowledge stored in the knowledge base. This view of quality for expert systems is related to the emulation of human experts since the knowledge base contains the expertise of the human expert.

As with conventional software, the quality of expert systems may be defined differently by the different people associated with the system. Many of the same types of people are involved during the development, operation, and maintenance of expert systems as are involved with conventional software. Expert systems, like conventional software, have a user, i.e., the person regularly working with the software in its operational environment, developer, and/or the person designing and constructing the software product, who is commonly known as a knowledge engineer, manager, and maintainer. The views of quality for these people are also often the same as those for conventional software. For example, for the users of an expert system, quality is not only the system's performance level but also the system's ability "to be fast, reliable, easy to use, easy to understand, and very forgiving when they make mistakes." [Waterman, 1986] A unique participant in the development of an expert system is the domain expert who provides the expertise to the expert system. For the domain expert, the quality of the expert system is the quality of the knowledge contained within the knowledge base. The quality of the reasoning process and the advice given by an expert system is also a concern for the domain expert.

The production of quality software for commercial use is a relatively new concept for expert system technology. As expert systems move further into commercial production, more will be researched and discussed related to defining quality for expert systems. This interest is already evidenced by the amount of recent work on enumerating the characteristics of quality expert system software.

III. CHARACTERISTICS OF QUALITY EXPERT SYSTEMS

Even in the earliest work on developing expert systems, the characteristics desirable in expert systems were examined. These characteristics, which in conventional software are considered as the quality attributes of software, are referred to by various names such as performance criteria, validation criteria, and evaluation criteria.

Gaschnig *et al.* [Gaschnig, 1983] provided some of the first evaluation criteria to describe the quality characteristics desirable in expert systems. This list of characteristics was the foundation for much of the evaluation in early expert system development; it has also been the basis for recent work in defining quality characteristics for expert systems. These evaluation criteria are:

1. The quality of the system's decisions and advice
2. The correctness of the reasoning techniques used
3. The quality of the human-computer interaction (both its content and the mechanical issues involved
4. The system's efficiency
5. Its cost-effectiveness [Gaschnig, 1983]

Much work has been done in recent years on further defining the criteria or characteristics by which the quality of expert systems is to be judged. Liebowitz [Liebowitz, 1986, Liebowitz, 1988] attempts to create criteria which span all aspects of expert system development by incorporating the work on quality characteristics of conventional software by Boehm et al. [Boehm, 1978] and the five criteria shown above defined by Gaschnig et al. [Gaschnig, 1983] These criteria are as follows:

1. ability to update — the capability of adapting software to changing system requirements
2. ease of use — understandability, user-friendliness, execution efficiency
3. hardware — portability, availability, and accessibility of the computer equipment needed
4. cost-effectiveness — costs and benefits involved in solving the task
5. discourse (input/output content) — explanation and help facilities
6. quality of decisions, advice, and performance — accuracy, consistency, and completeness of responses
7. design time — the length of time in person-years to solve the task. [Adapted from Liebowitz, 1986; Liebowitz, 1988]

The criteria of Boehm and Gaschnig are merged in each of Liebowitz's criteria. For example, the criterion "ease of use" incorporates Boehm's criterion of human-engineering and Gaschnig's criteria of the quality of the human/computer interaction and the system's efficiency.

Marcot [1987] addresses the evaluation of the knowledge base and provides an extensive list of validation criteria by which the validity of the knowledge base is to be evaluated. These criteria provide evaluation measures for the entire development process of the knowledge base. Appendix B presents this list of validation criteria which includes criteria which are specific to expert systems, e.g., breadth, depth, generality, and others which are similar to the characteristics of quality conventional software presented in Section III of Chapter 2, e.g., accuracy, reliability, robustness.

Guida and Spampinato [Guida, 1989a] have developed a comprehensive set of criteria for evaluating the adequacy of expert systems in critical domains. These criteria are designed for the assessment of the internal structure of the expert system during its development life cycle. The criteria are organized in a hierarchical manner with increasing detail on each successive level. The hierarchical decomposition of these criteria is shown in Figure 4.1. The top level in this set of criteria is composed of two characteristics which are:

1. The external **behavior** of an expert system, as it can be observed during operation
2. The internal **ontology** of an expert system, comprising its structural organization and content, as they result from design and construction [Guida, 1989a]

The criterion behavior is comprised of two characteristics: appropriateness and adequacy. Appropriateness deals with the results produced by the expert system. This characteristic is decomposed into coverage, i.e., the breadth of the domain handled by the expert system; granularity, i.e., the level of detail of the knowledge representations; capability, i.e., the diversity of problem situations which the expert system can handle; correctness and optimality, i.e., the optimality of the solution in relation to specified evaluation criteria or other solutions.

Adequacy measures how the expert system performs its problem-solving task. As seen on Figure 4.1, adequacy is comprised of eight characteristics. Uniformity is the consistency in behavior. Robustness is the ability to function near the boundaries of its coverage. Naturalness is the understandability to the user. Transparency is the explanation facilities. Effectiveness is the ability to function with available data. Efficiency is the use of computer resources. Friendliness is human-computer interaction.

The second major criterion, ontology, is decomposed into structure, content, software of the expert system, e.g., user inter-

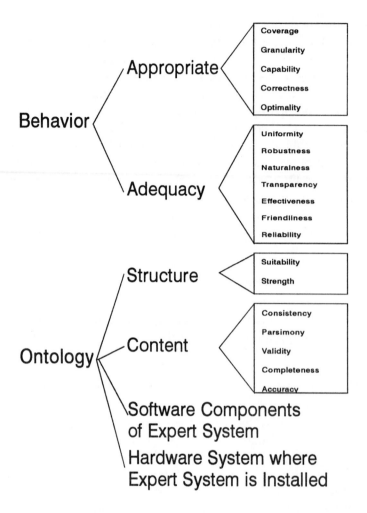

FIGURE 4.1.   Hierarchy of Evaluation Criteria

faces, reasoning processes, and hardware on which the expert system operates. Structure, which describes the type of knowledge representation or inferencing mechanism chosen for the domain of the expert system, is defined by the characteristics of suitability, i.e., appropriateness of the chosen knowledge representation or inferencing mechanism for the domain, and strength, i.e., measure of the use of good engineering principles in the development of the expert system. Content describes the structure of the knowledge base. This characteristic encompasses the characteristics of consistency, parsimony, i.e., the lack of unnecessary knowledge in the knowledge base, validity , completeness, and accuracy.

Guida and Spampinato suggest that these evaluation criteria can be used for verification and validation during the development of both critical and noncritical domains of expert systems. The purpose of these criteria is to provide measures for:

1. Assuring that the expert system is able to provide correct, possibly optimal, solutions to all problems of interest, and
2. Assuring that the expert system can behave such that the user can rely on its performance, being aware of its competence, confident in its ability and familiar with its limits [Guida, 1989a]

The authors do not address the issues of maintenance, e.g., modifiability, maintainability, or external considerations such as cost-effectiveness in this set of criteria.

More recently, research into quality criteria for expert systems has begun to address the quality of the development process as well as the quality of the expert system. Giarratano and Riley [Giarratano, 1989] developed a list of criteria encompassing features of both the development process and the software product. These criteria, referred to as metrics, include features such as maintainability, portability, and understandability of code which reflect how the software product was designed and implemented, i.e., its development process. The criteria are as follows:

1. Correct outputs given correct input
2. Complete output given correct input
3. Consistent output given the same input again
4. Reliable so that it does not crash (often) due to bugs
5. Usable for people and preferably user-friendly
6. Maintainable
7. Enhanceable
8. Validated to prove it satisfies the user's needs and requests
9. Tested to prove correctness and completeness
10. Cost-effective
11. Reusable code for other applications
12. Portable to other hardware/software environments
13. Interfaceable with other software
14. Understandable code
15. Accurate
16. Precise

17. Graceful degradation at the boundaries of knowledge
18. Embedded capability with other languages
19. Verified knowledge-base
20. Explanation facility [Giarratano, 1989]

Although usually not referred to as quality criteria, these evaluation, validation, and performance criteria describe the quality characteristics of expert system software. No one of these sets of criteria has been accepted as a standard for expert system development, with the exception of Gaschnig [1983] in the early development of expert systems. Work towards establishing a standard set of quality criteria will continue as long as expert systems are being proposed for commercial use.

IV. APPROACHES TO ACHIEVING QUALITY EXPERT SYSTEMS

Currently, the most common approach to evaluating the quality of expert systems has been the comparison of the recommendations or advice of the expert system with that of human experts. Typically, such evaluation has been performed in an informal, arbitrary fashion. The most formal evaluation of this type was conducted by the MYCIN developers who performed a blinded comparison with nine human experts.

The creation of development tools or environments for the production of expert systems is providing a means for improving the quality of expert systems. Examples of development tools include expert system shells and toolkits such as Knowledge Craft, ART, and KEE. None of the available development tools provide facilities for all aspects of the expert system development process.

Two popular areas of research related to the production of quality expert systems are knowledge acquisition and verification. The development of knowledge acquisition techniques and tools is to aid in the elicitation and organization of knowledge. Recent research into the verification of expert systems deals primarily with the verification of knowledge bases. The research in this area of verification is in developing tools for determining syntax and semantic errors in the knowledge base. This research has not been utilized on a widespread scale in producing commercial expert systems as yet.

Software engineering methodologies, software quality assurance, independent verification and validation, and configuration management are a few of the approaches to the assurance of quality conventional software which evolved as a result of the software crisis in the late 1960s. Although some in the expert systems field are beginning to acknowledge the need for such methods, techniques, or tools in the assurance of quality expert system software [Chapnick, 1990, Giarratano, 1989, Rushby, 1988], little has been done to incorporate or adapt these approaches to expert systems. Furthermore, the development of techniques and tools for providing comprehensive verification and validation of expert systems which are recognized as essential to the success of expert system technology in commercial applications is in its infancy.

5   EXPERT SYSTEM DEVELOPMENT

5   EXPERT SYSTEM DEVELOPMENT

I. INTRODUCTION

At the present time, expert system technology has no widely recognized and accepted methodology by which the development of expert systems is achieved. The most commonly used method of development is a trial-and-error method, called iterative refinement or prototyping. Simplistically stated, this method consists of obtaining a set of example test cases, determining from the domain expert how he/she would solve each of these test cases, and encoding this expertise into a knowledge base. This process is repeated until an expert system evolves which satisfies the research being conducted or which meets the customer's needs. When the iteration stops, the expert system is validated on a small set of test cases. These test cases are often the same ones upon which the knowledge base is built or scenarios encountered in the operational environment. In both situations, the verification and validation testing and the analysis of the results from the testing are performed in an informal, arbitrary manner. This testing and analysis entails the knowledge engineer or customer manually executing test cases and checking the results against the expected results. While this empirical approach to development was satisfactory for the development of expert systems within the research laboratory environment, more systematic, structured approaches to development are needed to produce quality expert systems for commercial applications. The empirical development of iterative refinement is "not supported by sound and general methodologies. It is more like handicraft than engineering, and it lacks several of the desirable features of an industrial process (reliability, repeatability, work-sharing, cost [estimation], quality assurance, etc.)." [Guida, 1989b]

Much work has been done in developing the methods or techniques needed to produce quality expert systems for commercial applications. The process model for expert system development, described in Section III below, illustrates many of the technical and managerial activities needed for the production of quality expert systems. Some of these activities are adapted from the concepts of the software engineering life cycle for conventional software. This model, like many recently proposed process models for expert system development, supports the belief that the "commercial development and implementation of expert systems necessitates a synthesis of the life-cycle and prototyping strategies as prototypes must eventually be transformed into usable production systems." [Agarwal, 1990] This process model represents the future in the development of expert systems; it does not represent common practice in expert system development. The common practice is iterative refinement. Research is still needed on many of the concepts presented in this example process model. Verification and validation are some of the concepts which require more definition, methods, techniques, and tools in order to be an integral part of the production of quality expert systems. The validation needed for expert system development, its relationship to the verification and evaluation of expert systems, and techniques for providing verification and validation of expert systems are described in the following sections.

II. VERIFICATION AND VALIDATION

Verification and validation are complementary approaches needed for assessing and assuring the quality of expert systems. Often in expert system terminology, verification and validation are associated with or are grouped under the topic of evaluation. The evaluation of expert systems is related to the criteria discussed in Section III of Chapter 4. Evaluation, sans verification and validation, deals with the utility of the system, i.e., those aspects which are especially relevant to the intended user(s) of the expert system; for example, the ease of use, efficiency, usefulness of the results, and cost effectiveness are examined. Evaluation activities deal with the usability of the expert system while verification and validation deal with the accuracy and reliability of the expert system.

Verification and validation activities occur throughout the development process for expert systems. Verification and validation are utilized together for:

1. Ascertaining what the system knows, does not know, or knows incorrectly
2. Ascertaining the level of expertise of the system
3. Determining if the system is based on a theory for decision making in the particular domain
4. Determining the reliability of the system [O'Leary, 1987]

For instance, verification and validation assess the consistency, correctness, and completeness of the knowledge within the expert system, the quality of the solutions provided by the expert system, and the ability of the system to produce the same results given the same inputs. In addition to these activities, verification and validation assure the following system requirements:

1. The compliance of the system with all the requirements established for it during the requirements development phase at the beginning of each prototype phase of the development cycle of the system. These requirements include input/output functional requirements, input/output performance requirements, and system/software compliance requirements.
2. The conformance of the system to the design from which it was built. The design used to build the system is usually picked from the set of feasible designs proposed for it. It also conforms to software, hardware, bioware, environmental and allocation of resources requirements.
3. The evaluation of the system by an expert, a group of experts, or an independent organization to determine the acceptance of the final system by the customer. [Jafar, 1989]

Verification examines the technical aspects of an expert system in order to determine whether the expert system is built correctly. Verification is "basically a test of whether the expert system matches the design ideas, i.e., whether it matches the technical requirements and expectations." [Hollnagel, 1989] Verification activities are performed by the developers of expert systems. One of the major tasks in verifying expert systems is the verification of the knowledge contained within the knowledge base. This type of verification assesses the accuracy of the knowledge. Inaccuracy in the knowledge base results in inaccuracy for the whole expert system. Verifying the knowledge base involves examining the consistency, correctness, and completeness of the knowledge by detecting errors such as redundancy, contradiction, and circular dependency. Another major task of verification is verifying the decision-making functionality of the expert system. The decision-making functionality is verified by examining the inference engine and the reasoning process of the system. Verification of the decision-making functionality assesses not only if the expert system is producing correct intermediate and final results but also if the expert system is using the correct reasoning process while determining the correct results. Verification is accomplished utilizing static and dynamic analysis.

Static analysis, which does not involve the execution of the expert system, is utilized, for example, to verify the knowledge base. Static analysis of the knowledge base is the manual or automated analysis of the contents of the knowledge base in order to determine the consistency, completeness, and correctness of the knowledge. Dynamic analysis, which involves the execution of the expert system, is also used in verification of expert systems. For example, determining if the expert system is producing the correct answers and using the correct reasoning process is a verification activity performed by the dynamic analysis of the expert system.

Validation is used to determine whether the expert system meets the needs and expectations of the customer. Validation substantiates that "the expert system performs the desired task with a sufficient level of expertise." [Prerau, 1990] Validation for expert systems is composed of two types of validation — validation of explicit requirements and validation of implicit requirements.

The explicit requirements for expert systems are the functional and performance requirements which are delineated in a requirements specifications document which is produced early in the development of an expert system. These requirements, which the customer has explicitly specified, are to be demonstrated in the expert system in order for it to be acceptable to the customer. Since this type of validation is the same as validation in conventional software, all the techniques and tools used there can also be applied to this type of validation for expert systems. At the present time, expert system developers have only begun to address the issue of explicit requirements. This type of validation testing has not been implemented in common practice. In fact, the ability to delineate such requirements for an expert system is still a debated issue among expert system developers and is discussed in further detail in Section III of Chapter 6.

The implicit requirement to be validated is reflected in the definitions of quality for expert systems discussed in Section II of Chapter 4. Performance equivalent to the performance of a human expert is a requirement implicit to all expert systems. If the performance of the expert system is not required to be equivalent to the performance of the human expert, the level of performance, acceptable to the customer, must be defined explicitly. This implicit requirement is unique to expert systems; therefore, the validation of this requirement is also unique. Furthermore, the validation of such an elusive requirement must be systematic and well documented so as to convince the customer of the successful realization of this requirement. To demonstrate the satisfaction of this implicit requirement, expert systems are validated against human experts. This type of validation deals with measuring the level of expertise at which the expert system performs. This measurement is usually accomplished by comparing the performance of the expert system with the performance of a human expert on a set of inputs from or similar to real world situations. The level of performance achieved by the expert system is expected to be equivalent to that of the human expert or equivalent to a predefined level of acceptance. Only the quality of the final decisions of the expert system are considered in this type of validation of expert systems. Dynamic analysis is utilized to validate both the implicit and the explicit requirements.

In the next section, more description of the verification and validation activities needed for the development of quality expert systems is provided within the framework of the Linear Model process model.

III. OVERVIEW OF EXPERT SYSTEM PROCESS MODELS

Several versions of process models have been proposed for the development of expert systems. The following process model

[Giarrantano, 1989] is representative of the latest work on developing systematic approaches for the commercial production of quality expert systems. This process model, named the Linear Model, incorporates activities unique to expert systems, e.g., accounts for iterative development, with tasks necessary for quality commercial production, e.g., provides system documentation, verification, validation. The phases, the reviews, and the products of the phases for the Linear Model are shown in Figure 5.1. The phases of the Linear Model, which are planning, knowledge

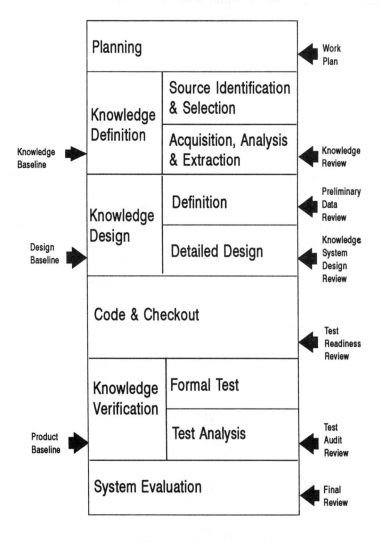

FIGURE 5.1.   The Linear Model

definition, knowledge design, code and checkout, knowledge verification, and system evaluation, occur in each major iteration in the development of an expert system. The typical major iterations in the development of quality commercial expert systems are shown in Figure 5.2. The functionality, verification activities, and validation activities for each phase of the Linear Model are described in the following sections.

A. Planning Phase

The major activities in the first phase of the Linear Model are a feasibility study and an initial analysis of the problem and its development. The appropriateness of the problem for expert

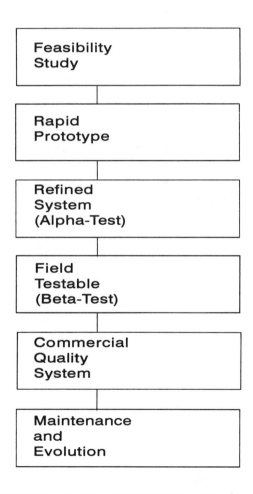

FIGURE 5.2. Iterations in Expert System Development

system development and its cost-effectiveness are examined in the feasibility study. The initial analysis of the problem and its development includes determining the resources of hardware, software, and personnel needed to build the expert system, delineating the tasks for solving the problem, creating a time schedule for these tasks, and describing the high-level functional requirements for the proposed system. The information gathered from these activities is compiled in a document which is called the work plan. This work plan provides the guidelines for the rest of the development process.

B. Knowledge Definition Phase

The two main activities in the knowledge definition phase are the knowledge source identification and selection and the knowledge acquisition, analysis, and extraction. The knowledge source identification and selection activity includes the identification of knowledge sources, e.g., human experts, books, manuals, and the prioritization of this list of knowledge sources. These two activities are completed without regard to the availability of the knowledge sources. The knowledge sources are then ranked by availability, and the knowledge sources to be utilized during development are selected.

The second main activity in the knowledge definition phase is the acquisition, analysis, and extraction of knowledge. The output of these activities is a set of knowledge which has been reviewed, verified, and baselined. Other activities of this phase include the description of the functional capabilities of the system, production of a preliminary user manual, and definition of requirements specifications.

The verification activities during this phase are the verification and formal review of the knowledge. The review process results in the knowledge being baselined, i.e., put under the control of configuration management. The definition of the requirements specifications is a validation activity. These requirements specifications, which are the basis for validation testing, describe the expected functionality of the expert system.

C. Knowledge Design Phase

The knowledge design phase of the Linear Model constructs the detailed design of the components in the expert system. The design of the knowledge representation, the inference engine, and the user interface are performed during this phase. The activities which encompass the production of this detailed design are decomposed into a knowledge definition activity and a detailed design activity. The preliminary work produced in the knowledge definition activity is reviewed before the detailed design begins. Also produced during this phase is the test plan which acts as guidelines for the testing of the expert system. The delineation of test data, the specification of test drivers, and the description of the documentation and analysis of test results are included in the test plan. All the decisions and information resultant of this phase are documented in a design document.

The verification and validation activities within this phase include the review of the preliminary design information, the formal review of the design document, and the generation of the test plan. The formal review of the design document results in the baselining of this document.

D. Code and Checkout

The activities of the code and checkout phase include the implementation and testing of the code as specified in the design, the documentation of source code, the production of the final version of the user manual, the generation of the installation/operation manual, and the documentation of the functionality, limitations, and problems of the expert system. A formal review of the readiness of the code for testing and the documentation produced here is conducted at the end of this phase.

E. Knowledge Verification

The knowledge verification phase of the Linear Model is the testing phase of the expert system development process. The testing activities specified in the test plan are implemented, the results are documented and analyzed, and conclusions and suggestions are made based on these results. The implementation and documentation of the tests comprise the formal test activity of the knowledge verification phase. The analysis and generation of suggestions comprise the test analysis activity of this phase. A review of the test documentation is conducted. The software product, i.e., the expert system produced in the code and checkout phase and tested here is baselined in a formal review at the end of the knowledge verification phase.

F. System Evaluation

The primary activities in the system evaluation phase are the evaluation and validation of the expert system. The evaluation of the expert system is based on the results of the testing performed in the previous phase (as reported in the test documentation). This evaluation process involves both the examination of these testing results and the recommendation of modifications to the expert system. The validation testing of the expert system, which is also performed during this phase, is designed to demonstrate that the expert system correctly implements the user's requirements. A report on the conclusions of the evaluation and validation processes is produced and reviewed. The report addresses questions such as:

1. Does the system meet originally defined objectives and standards?
2. Can it be used in the intended problem domain within an acceptable error rate?
3. Can it be successfully placed in the market?
4. What are the needs for adding to or revising the knowledge base?
5. Are there additional considerations to address, such as legal or regulatory constraints?
6. What can be learned to help in the development of the next system? [Marcot, 1987]

This report, along with all baselined items and documentation produced by the activities of the Linear Model process model, is either the input to the next major iteration of the development process or the final report on the expert system.

IV. RESULTS OF USAGE

No information exists on the success or failure of systematic approaches to expert system development as shown in the Linear Model. These types of process model are only beginning to be used in the development of expert systems. Little is known about the success or failure of iterative refinement since very few expert systems reached commercial production using this method. One of the systems which has proven successful in industry is the R1/ XCON developed at Carnegie-Mellon University for Digital Equipment Corporation (DEC). The R1/XCON, which configures VAX computer systems, was validated on 50 test cases. After its first year in use, it was observed that DEC personnel were examining the configurations recommended by the R1/XCON and modifying 40 to 50 percent of these recommendations. Furthermore, it was questionable whether the recommendations of R1/XCON were actually used by the DEC technicians who constructed the VAX systems. [Rushby, 1988] However, with the reimplementation of the system to correct these problems, the R1/XCON has become a financial success. DEC claims that this expert system saves the company $40 million a year. [Blackman, 1990]

V. SUMMARY OF VALIDATION IN EXPERT SYSTEM DEVELOPMENT

The validation activities, reviewed in this section, are illustrated in the Linear Model or discussed in other work on the validation of expert systems. The current practices in iterative refinement are also described where applicable. Although more systematic approaches such as the Linear Model are addressing problems encountered in early expert systems, precise details on how to accomplish some of these suggested activities are undefined. The details of validation not specified in the Linear Model, e.g., types of test cases needed, techniques for generating test cases, are examined in Sections IV and V of Chapter 6.

A. Validation Testing

With iterative refinement no distinction is delineated between the testing activities of verification and validation. The primary testing of the expert system occurs at the end of the development process in the form of validation testing. This type of testing is to compare the system's performance with that of a human expert. As stated earlier, much of this testing is performed in an informal, arbitrary fashion.

In more systematic approaches such as the Linear Model, the distinction between verification and validation is becoming clearer. More formality is added to the validation process. The validation process is composed of the validation of the functional requirements explicitly stated in the requirements specifications and the validation of the equivalence in performance to human expert(s). Functional requirements validation is the same as the validation of conventional software, and the techniques and tools discussed in Section V of Chapter 3 are applicable here. The validation of the equivalence to human expert(s) is unique to expert systems and is addressed in the remainder of this section.

Validation of the expert system's performance against the human expert's performance is performed by the dynamic analysis of the expert system using test cases and/or live tests. Such validation testing is divided into quantitative validation and qualitative validation. Quantitative validation utilizes statistical methods to compare the expert system performance against the human expert performance. Qualitative validation utilizes subjective measures in comparing the expert system performance against the human expert performance. Common approaches to qualitative validation include face validation, predictive validation, Turing tests, and field tests.

Face validation is a preliminary approach to validating an expert system. In face validation, a cursory appraisal of the expert system's performance is conducted by the development team, domain experts, and potential users to assess the "consistency between the system designer's view and the expert's view of the expert system in a timely and cost-effective manner." [O'Leary, 1990] Face validation allows all parties involved to assess whether a satisfactory solution is possible for the specific problem.

Predictive validation is the use of historic test cases for the validation of expert systems. These historic test cases have either known results or measures of human expert performance for those test cases. Therefore, the expert system is executed with these test cases, and its results are compared with the known results or measures. This approach is also known as backcasting.

The most commonly used method of validation testing is the Turing test. In the Turing test, a blind comparison is made between the expert system's results and the human expert's results on a set of test cases. The blind comparison is achieved by placing both sets of results into a standard format so that the origin of the results is undetectable. A judge or group of judges then assesses the results. If there is no distinguishability between the sets of results, the expert system has accurately emulated the human expert. The Turing test is important in that it eliminates possible bias for or against computers in the validation of expert systems. In the validation of MYCIN, the initial comparison was not blinded, and the biases of the judges against computers were reflected in their evaluations of the performance of the expert system. Another comparison was performed where the origin of the results was blinded, and the biases of the judges were no longer a factor in the validation process. [Gaschnig, 1983]

In field testing, the expert system is placed in its operational environment, and the user performs live testing which acts as the validation of the expert system. Field testing is only suitable for the validation of noncritical applications. This type of testing is usually very informally conducted. An advantage for the knowledge engineer is that the user assumes the task of validation testing. [O'Keefe, 1987] Furthermore, successful field testing increases customer acceptance more than laboratory testing. [Prerau, 1990]

B. Validation Test Cases

Test cases are an integral part of the development of expert systems. Test cases are used to elicit knowledge from the domain expert about how problems are solved, i.e., knowledge acquisition, and to verify and validate the expert system. The availability of test cases is domain dependent. Some domains have easily obtainable test cases, and others do not. Sources for test cases are real world situations, e.g., historical files, ongoing transactions, and manual generation by the domain expert or knowledge engineer.

In iterative refinement, the number of test cases used in the validation of expert systems typically has been very small, and often the test cases used for knowledge acquisition were also used in validation. In the Linear Model process model, test cases are part of the test plan which acts as the guidelines for the verification and validation testing activities.

C. Acceptance Criteria

In the Linear Model, the requirements specifications contain the functional and performance requirements for the expert system being developed. These functional and performance requirements are the acceptance criteria of the expert system. The acceptance criteria are the features of the expert system which are to be demonstrated to the customer in order to declare the system as successful. As seen in the Linear Model, these acceptance criteria are established early in the development process in order to prevent misunderstanding between the customer and the developers which would lead to the development of unwanted and unnecessary features in the expert system.

D. Testing Documentation

As seen in the Linear Model, test plans are the only type of testing documentation to be considered for expert systems thus far. The test plan acts as the guidelines for the testing activities of verification and validation. A test plan includes a statement of purpose which describes the objectives of the testing and the criteria for success of the testing, the approach to be used in the testing, the personnel and facilities required, a schedule for the testing process, and procedures for the documentation and analysis of test results. A test plan also contains the test cases and the expected results to be used in the testing process.

E. Revalidation During Maintenance Phase

Revalidation, which is also known as regression testing, is the testing of the expert system during the maintenance phase. This type of testing occurs after modification is implemented in the expert system. Revalidation ensures that the quality of the original expert system remains in spite of changes introduced to the system and that the modifications are acceptable. Very few expert systems have endured to the maintenance phase thus far. Although little research has been done related to the maintenance of expert systems, maintenance is an issue to be addressed if expert systems are to be successful as commercial applications. Additionally, with the nature of an expert's knowledge being one of constant growth and change, the maintenance of expert systems will require techniques and tools for handling change within the knowledge base.

The most proposed means of revalidating expert systems is the application of a set of test cases which were successfully executed on the expert system before modification. Comparison with the performance from previous executions with the performance after modification(s) indicates the effect of the modification(s) on the expert system's performance level. New test cases which exercise the modifications are also added to the set of test cases.

F. Validation Tools

Currently, no tools exist which automate the activities of validation for expert systems. The tool presented in this book will be one of the first developed. There has been much interest and work done in developing tools to automate the activities of verification. Most of these tools center around the syntax and semantics of the knowledge base. These tools are designed to replace the manual examination of the knowledge base as the primary form of verification. These tools examine the knowledge base for problems such as redundancy, contradiction, circular dependency, unused or undefined items, and unreachable items. [Stachowitz, 1987; Nguyen, 1987; Kang, 1990]

6

ISSUES AND RECOMMENDATIONS IN EXPERT SYSTEM VERIFICATION AND VALIDATION

ISSUES AND RECOMMENDATIONS IN EXPERT SYSTEM VERIFICATION AND VALIDATION

I. WEAKNESSES OF CURRENT PRACTICE

The expert system verification and validation, as described in Section I of Chapter 5, has several weaknesses. Misinterpretation of verification and validation testing, random selection of test cases, and manual comparison of execution results are some of the weaknesses which make the verification and validation process informal, subjective, time-consuming, tedious, and arbitrary.

A major weakness of the current approach to the verification and validation of expert systems is the misinterpretation of verification and validation. The testing in this approach is a mixture of verification and validation. Verification and validation are complementary, but distinct approaches to the assessment and assurance of quality software. Each approach has its own set of objectives and activities for achieving these objectives. The testing in the current approach to verification and validation has no distinct objective. The objective, which is arbitrarily that of verification or of validation, is dictated by the nature of a specific test case or by the test administrator, i.e., the person performing the execution of the test case. The test administrator is generally the knowledge engineer or the customer. However, both the knowledge engineer and the customer have distinct viewpoints of the expert system and distinct objectives in testing. Using one arbitrarily over the other as the test administrator affects the consistency and effectiveness of the verification and validation process.

Another weakness of the current approach to verification and validation testing is the number and selection of test cases used. Typically, the number has been very small; for example, only 10 test cases were utilized for the validation of MYCIN, and 50 for R1. Additionally, the selection of the test cases has been unsystematic. Test cases are chosen without regard to their representation of the problem or knowledge domain. The selection process also does not consider the coverage of test cases. Coverage is the measurement of how effectively the test cases verify or validate the expert system.

An additional weakness in the current practice of expert system verification and validation is the manual execution of test cases and the manual comparison of results. Such manual activities are time-consuming and tedious. Currently, there are no tools to alleviate this problem for the verification and validation of expert systems.

The use of a human judge in the manual comparison of results is another weakness in current expert system verification and validation. A human judge compares the results of the expert system with the results of human expert(s) and decides on the equivalency of these results. This decision is subjective since it may vary depending on the judge. Human judgement is affected by personal bias and parochialism. Bias for or against computers may influence the judge's decision; however, this bias can be eliminated with the use of a blinded evaluation, e.g., Turing test. Parochialism is another factor in the judge's decision. Parochialism is the effect of the region of the country or world, in which the judge lives or was educated, on his/her evaluation criteria or knowledge. [Green, 1987] Moreover, the use of a judge in verification and validation is difficult since the availability of someone knowledgeable enough to perform such a judgement in a particular domain is often limited. The only people available may be the domain expert or the knowledge engineer from the development process; this situation is not desirable because of their lack of objectivity towards the knowledge which they generated or the expert system which they developed. The source for the results of test cases against which the expert system's results are compared is another weakness in the comparison of results. Because of the limited availability of domain experts, the expert system is often compared with the human expert who acted as the domain expert during the development process. Such validation, which was the case with the validation of the expert system Prospector, should be viewed skeptically. [O'Keefe, 1987]

Other weaknesses of the current approach to expert system verification and validation include the lack of acceptance criteria and requirements specifications, the lack of quantitative assessment techniques, and limited testing documentation. These issues are addressed in the succeeding sections.

II. INSIGHTS FROM PAST EXPERIENCE

Past verification and validation experience is a source of insights about problems and solutions in the validation of expert systems. Gaschnig et al. [1983] delineated seven suggestions for the evaluation of expert systems which are applicable to the verification and validation process. In these suggestions, the need for verification and validation planning was emphasized. Three suggested planning activities were designating for whom the validation activity is intended, e.g., the development team, customer, defining precisely what is being validated, and identifying who is performing the validation activity. These planning activities differentiate the objective of validation testing for expert systems. Additionally, identifying the type of test cases to be used in validation, e.g., random test cases, test cases selected for a specific purpose, was recognized in these suggestions as being essential to evaluating the type of coverage achieved in validation. Other suggestions by Gaschnig et al. included eliminating any bias in the validation process and choosing a realistic measure of performance against which the expert system is validated. Usually, this realistic measure of performance is not against perfection but is "against the proficiency of typical domain practitioners." [Prerau, 1990] Finally, specifying a "gold standard" against which the performance of the expert system is compared was suggested for the validation process. This "gold standard" can be "the objectively correct answer to the problem" or "the answer given by a human expert, or group of experts, when presented with the same information as the system." [Rushby, 1988]

In addition to the suggestions described above, Gaschnig et al. [1983] also recommended that planning for evaluation of the expert system occur as part of the system design process at the beginning of the development of the expert system before system is built. Part of this planning includes specifying the objectives of the expert system and "explicit statements of what the measures of the program's success will be and how that failure or success will be evaluated." [Gaschnig, 1983] Furthermore, it was advised that evaluation be an integral part of the development process of an expert system and be associated in the various phases of development. All aspects of these recommendations for evaluation are also applicable to validation. Finally, Gaschnig et al. emphasized that the customer needs to be involved in the acceptance testing of the expert system.

Other lessons learned from past verification and validation experiences include the need for using both qualitative approaches, e.g., Turing test, field testing, and quantitative approaches in the validation of expert systems. [O'Keefe, 1987] Better selection of test cases and clear delineation between development testing and acceptance testing are two other lessons learned. [Rushby, 1988]

All of these insights from past verification and validation experience present problems and solutions which are important considerations for the verification and validation of expert systems. The recommendations presented in this book include techniques and tools for resolving many of these problems or for implementing the suggested solutions.

III. DIFFICULTIES IN VERIFICATION AND VALIDATION

The lack of requirements specifications and acceptance criteria is a major difficulty encountered in the validation of expert systems. As stated earlier, requirements specifications are necessary elements in the validation process. Without a specification containing acceptance criteria, the concept of acceptance for the expert system is unclear and is subject to misinterpretation. [St. Johanser, 1986] However, expert system developers have been reluctant to undertake the task of developing requirements specifications for the systems which they build. The ill-defined nature of the problems which expert systems address is often perceived as the reason for the inability to develop requirements specifications. The research environment in which expert systems were typically developed in the past and the small size of the projects also allowed for the lack of requirements specifications. Other reasons include:

1. The user and developer plan to begin prototyping from a statement of purpose and refine the requirements during prototyping.
2. The user or the developer is unwilling to commit the resources necessary to prepare and maintain a requirements specification.
3. The developer believes that imposing detailed documented requirements would unduly constrain the developer's creativity.
4. The user does not plan to perform a formal acceptance test. [Green, 1987]

The need for quality assurance techniques such as validation for expert system technology to receive widespread acceptance in the commercial computing community has already been discussed in Section II of Chapter 1. Furthermore, the use of quality assurance techniques in the development of expert systems necessitates precise requirements and specifications. [Rushby, 1988] The hazards of omitting the development of requirements specifications were experienced by conventional software developers. Expert system developers and users are beginning to recognize these same problems in expert system development: "Rapid prototyping may help build systems quickly, but building an incorrect or incomplete one 10 times faster is no gain." [Chapnick, 1990]

Many problems, which other areas of artificial intelligence undertake, are such that the development of requirements specifications is difficult. However, because expert systems are more problem-oriented than process-oriented, like other AI applications, expert systems are more amenable to the definition of requirements specifications. [Geissman, 1988] Embedded expert systems are especially accessible to requirements specifications because of the exacting operational environment in which they must exist. Furthermore, research has begun in defining categories of requirements which are reflective of the uniqueness of expert systems. [Rushby, 1988]

Validating expert systems presents several other difficulties. The type of results which expert systems produce makes validation of expert systems complicated. For conventional software, results produced during validation testing are easily validated because they are exact in their correctness, i.e., the results are either correct or incorrect. It is harder to judge the results of an expert system in such an exact manner. Validating an expert system is to validating traditional software as "grading an essay examination is to grading a true-false examination." [Green, 1987] Not only do expert systems often produce complex results, but it is also difficult to have human experts agree upon what should be the correct result. There may be several acceptable results for a particular problem or no agreement among several experts on what the acceptable result actually is. Expectations on the level of expertise to be demonstrated by an expert system also differ among human experts.

IV. OVERVIEW OF RECOMMENDATIONS IN PROCESS MODEL

Where applicable, the concepts of verification and validation in conventional software, described in Chapter 3, are adapted for expert system verification and validation in the recommendations presented in this book. The verification and validation activities for expert system development, presented in Chapter 5, and the lessons learned from past verification and validation experiences, described above in Section II, are also incorporated into these recommendations. Moreover, new techniques or tools for validation of rule-based expert systems are presented where the uniqueness of rule-based expert systems demands them and where current practices are inadequate.

In this section, an overview of the recommendations of this book is presented. These recommendations are described within the framework of the Linear Model. Where necessary, the verification and validation activities of the Linear Model are more clearly defined or are modified to accommodate the recommendations of this book.

The recommended verification activities are summarized in Figure 6.1, and the recommended validation activities are summarized in Figure 6.2. The recommendations, which are delineated in the following sections, are discussed further in Section V.

Planning		Generation of preliminary Project V & V Plan
Knowledge Definition	Source Identification & Selection	Verification of knowledge Formal review of knowledge Collection of test cases
	Acquisition, Analysis & Extraction	
Knowledge Design	Definition	Review of preliminary design
	Detailed Design	Formal review of design document Generation of multiple test plans Completion of Project V & V Plan Completion of test case collection
Code & Checkout		Formal review of code and other documentation produced here Static verification of knowledge base Unit test Generation of test cases
Knowledge Verification	Formal Test	Implementation of verification tests Generation of test cases
	Test Analysis	Documentation of tests Review of test documentation
System Evaluation		

FIGURE 6.1. Recommended Verification in Linear Model

A. Planning Phase

The production of a preliminary Project V & V Plan is recommended. This documentation describes a systematic, comprehensive plan for the verification and validation of the expert system under development. The Project V & V Plan is completed, reviewed, and baselined during the next phase.

B. Knowledge Definition Phase

The output of the knowledge definition phase in the Linear Model is a set of knowledge which has been verified, reviewed, and baselined. Since the knowledge is not encoded into the knowledge base until a later phase of the Linear Model, the only type of verification possible in this phase is the manual examination of the knowledge. This manual examination can be successfully accomplished using the techniques of walkthrough or inspection as described in Section IV of Chapter 2.

Planning		Generation of preliminary Project V & V Plan
Knowledge Definition	Source Identification & Selection	Definition of requirements specification Collection of test cases
	Acquisition, Analysis & Extraction	
Knowledge Design	Definition	Generation of multiple test plans Completion of Project V & V Plan Completion of test case collection
	Detailed Design	
Code & Checkout		
Knowledge Verification	Formal Test	Validation by development team Documentation of tests Generation of test cases
	Test Analysis	
System Evaluation		Validation by customer Documentation of tests Generation of test cases

FIGURE 6.2.   Recommended Validation in Linear Model

The definition of requirement specifications is part of the validation activities of the knowledge definition phase, as specified in the description of the Linear Model. The method for defining these requirement specifications is beyond the scope of this book, but the need for such documentation has been discussed in Section III. The requirements specification document must delineate the functional and performance requirements for the expert system in unambiguous, testable statements upon which validation testing can be based. The requirements specifications document is reviewed and baselined during this phase of the process model.

It is recommended that the collection of test cases be begun during the knowledge definition phase. The personnel, specified in the Project V & V Plan as being responsible for the collection of test data or for the administration of particular verification and validation testing activities, coordinate the collection of test cases. The designated collector of test data collects test cases from real world scenarios, historical cases, or on-going transactions. Good sources for such test cases are the domain expert and the customer. The availability of such test cases is domain dependent. The intended use of the test case must be specified, and the test cases are maintained in an automated repository for that specific use. Test cases for validation are collected for both the implicit and explicit requirements.

C. Knowledge Design Phase

The primary verification and validation activities in the knowledge design phase are informal and formal reviews of the design document and generation of a test plan. Other verification and validation activities are needed during the knowledge design phase.

The Project V & V Plan, which is produced in the knowledge definition phase, is examined in regard to the design document. Modifications and additions may be required based on the design decisions. If any modifications or additions are necessary, the Project V & V Plan is reviewed and baselined again.

A test plan for each type of verification and validation testing, specified in the Project V & V Plan, is produced. The test cases for the various verification and validation testing activities are included in these test plans. The collection of these test cases was begun in the previous phase and is completed here. The different test plans are reviewed and baselined along with the design document, but these test plans remain as separate entities and are not incorporated into the design document.

D. Code and Checkout

The verification activities during the code and checkout phase are the formal review of the test readiness of the code, the review of the documentation produced in this phase, and the testing of the code. No description of the type of testing to be performed in the code and checkout phase is supplied by the Linear Model.

The most appropriate type of testing at this phase of the development process is the dynamic verification activity of unit testing. Expert system developers define unit testing as "incrementally testing each new feature or major addition to the knowledge base." [Irgon, 1990] This testing is performed according to the test plan for unit testing.

Automated verification tools such as EVA [Stachowitz, 1987], ARC [Nguyen, 1987], and VALIDATOR [Jafar, 1989], which perform static verification of the knowledge base, are also applied at this point in the development process. Such static verification tools are designed for the syntactic and semantic analysis of the contents of the knowledge base. These verification tools, which are primarily designed for rule-based systems, check for errors such as rule redundancy, effect conflict, condition conflict, circular dependency, subsumption, unnecessary conditions, and unreachable conditions. If no such tools are available, the syntactic and semantic analysis of the knowledge base must be performed manually by the domain expert.

E. Knowledge Verification

The testing activities specified in the test plan are implemented, the results are documented and analyzed, and the software product is formally reviewed and baselined during this phase. The exact nature of the testing performed during the knowledge verification phase is vague in the Linear Model description. This testing is defined as verification and validation in this book.

At this point in the development process, because of previous verification activities, the knowledge base should be free from internal errors. The verification objective, therefore, is to test the parts of the expert system operating together, either incrementally or all at once. This verification testing examines the implementation of the knowledge, i.e., the answers of the expert system and the reasoning process used to produce the answers. Verification of the inference engine and user interface is unnecessary with the use of commercial development tools such as expert system shells. After the development team is reasonably sure that the expert system is operating correctly, i.e., the verification objectives have been satisfied, validation testing begins.

Validation during development examines the level of expertise and accuracy of the expert system, i.e., the implicit requirement, and the explicit requirements of the expert system. Validation demonstrates to the developers whether the expert system is ready for validation by the customer. The use of an automated tool to assist in this validation testing and in the comparison of the testing results is recommended. The automated comparison of testing results also applies statistical measures in order to determine the equivalency between the results of the expert system and the results of human expert(s).

The validation testing in the knowledge verification phase is performed by the development team in the laboratory environment. The testing follows the guidelines of the Project V & V Plan and the appropriate test plan. The test cases in the validation repository are utilized. The effectiveness of the test cases in the repository is analyzed. More test cases may be needed in order to more thoroughly validate the expert system. For example, test cases may be needed to more completely validate the rules within the knowledge base. An automated tool to perform this coverage analysis and to generate suggestions for test cases is recommended. When the expert system is released from this phase of development, the development team is confident in its reliability and accuracy and in the presence of the functionality specified by the implicit and explicit requirements.

F. System Evaluation

Validation during the system evaluation phase, as described in the Linear Model, is the determination of the correctness of the expert system with the user's requirements. The implementation of this validation is undefined in the Linear Model. In the recommendations presented here, at this point in the development process, the development team has already performed validation testing and is satisfied that the functionality, displayed by the expert system, meets the needs and requirements of the user. Therefore, the validation during the system evaluation phase is the validation for customer acceptance. The purpose of validation for customer acceptance is to demonstrate to the customer (user) that the expert system performs the required functionality in a reliable fashion.

Validation for customer acceptance uses test cases which are stored in the validation repository and which exercise both the implicit and explicit requirements of the expert system. This validation testing is performed by the customer or the customer's representative in the operational environment of the expert system. The customer should have access to the same automated validation tools described for the validation during development.

V. VERIFICATION AND VALIDATION
RECOMMENDATIONS

The recommendations presented in this book provide a more formal, objective, and automated means of validating rule-based expert systems. Adding formalism to the validation process "(1) establishes when validation should occur within the development life cycle, and (2) identifies validation methods, input domain specification, the level of acceptance, and (where appropriate) the relevant application of statistical techniques." [O'Keefe, 1987] Objectivity, achieved through the use of statistics measures, gives more credibility and reliability to the validation process. Automation in the validation process promotes effective use of the time of domain experts and eliminates the time-consuming, tedious tasks of validation.

A. Validation Testing

Validation activities occur throughout the development life cycle of expert systems. The three occurrences of validation testing for expert systems are during development, for customer acceptance, and during operational use. All three types of validation address the testing of the implicit and explicit requirements of the expert system.

Validation during development is administered by the development team in the laboratory environment, i.e., non-operational environment. The purpose of validation during development is to demonstrate the level of expertise and accuracy of the expert system to the development team and possibly management. This validation testing also demonstrates that the functionality, specified by the customer in the requirements specification document, is satisfactorily implemented in the expert system.

After the development team has completed its validation testing and is satisfied with the results, the expert system undergoes validation for customer acceptance. The purpose of this validation testing is to ensure acceptance of the expert system by the customer. Validation for customer acceptance is ideally performed in the operational environment of the expert system. If the application is noncritical, validation for customer acceptance may be accomplished through the technique of field testing. If the application is critical, the operational environment of the expert system must be simulated. The development team may need to guide the customer in this validation testing in order to have formal, effective testing rather than just tinkering with the expert system. The test plan generated for validation for customer acceptance acts as the guidelines for the customer's responsibilities in this activity. Some expert system developers suggest that this testing be used like the acceptance testing of conventional software where "the user formally signs off on the quality of the decisions made by the system." [O'Leary, 1987] The criticality of the application or the concern for legal liability may require this formalism.

Validation continues after the expert system is in operational use, i.e., maintenance. Validation during operational use is known as revalidation. During the operational use of the expert system, enhancements or modifications may be implemented. Such changes are especially true of an expert system since the knowledge of an expert system is constantly changing and evolving. After the implementation of any changes to the expert system, revalidation ensures that the level of expertise and accuracy of the expert system is the same as or better than before the changes and that none of the required functionality or performance of the expert system has been lost or degraded. Revalidation is described further in Section V of this chapter.

Validation during development, validation for customer acceptance, and revalidation are performed through the execution of a set of test cases by the black-box technique of conventional software. The answers of the expert system on these test cases are compared with the answers of human expert(s) on the same test cases, or with known answers in the use of historical test cases. This comparison, as seen in Section III, is a difficult process with the possibility of multiple experts, multiple answers, and certainty factors. Determining a clear success or failure is often problematic. The use of statistical measures in this comparison is recommended as a means of evaluating the equivalency of the two sets of answers. Such measures include order compatibility and distance comparability, which are described in Section II in Chapter 7. The use of a tool which automatically performs the execution of test cases and the comparison of results is also suggested.

Validation, however, is more than the execution of test cases. The preliminary activities of validation include defining the objectives of testing and performance criteria, specifying the resources and personnel needed, and determining the schedule of testing activities. The test cases to be used in the validation process and their expected results must be collected or generated. The method of accumulating and analyzing the results from validation testing must be established. This preliminary validation work is accomplished through the preparation of the Project V & V Plan and the test plans for validation which are discussed in Section V.

B. Validation Test Cases

Test cases, utilized in expert system development, have distinct functions. These functions are knowledge acquisition, verification, and validation. The test cases of knowledge acquisition should not be used for verification and validation since these are the test cases upon which the expert system is built. Verification test cases should not be used for validation since the expert system has been tailored to operate properly for these test cases. The typical types of test cases in expert system development are real-world scenarios, historical cases, on-going transactions, and generated cases. Generated cases are manually developed by the system developer or domain expert to provide test cases when the number of real-world scenarios, historical cases, or on-going transactions is insufficient.

The effectiveness of test cases utilized in validation is important. It is impossible to produce a set of test cases which handles every combination of input which the expert system may encounter; therefore, a representative set of test cases must be used to validate the system. [Hall, 1988] The following suggested criteria for test case selection, based on the test case generation techniques of conventional software, illustrate the many considerations in developing a representative set of test cases for validating an expert system:

- Test every requirement. For each requirement, there must be a test case capable of showing that the requirement has been met. (This is why there must be a requirements specification that contains only testable requirements.)
- A string of test cases that repeatedly pushes the system against one of its limits or causes error-recovery or housekeeping activities to be invoked is often more effective at revealing errors than a set of independent test cases.
- Test every item of code and every possible decision. Equivalence partitioning may be used to keep the number of test cases reasonable. For expert systems this means that every

fact, object, rule, etc. must be invoked, every line of every method must be run, and every possible output of each rule must be caused. Particularly in expert systems, where the flow of processing is not always clear, diligent study may be required to determine how to sensitize a particular decision.

- In expert systems it is necessary to test both the outputs of the system and the process by which the system arrived at those outputs.
- Test every singular point and boundary condition. Test cases just below, right on, and just above a boundary are needed to characterize behavior around the boundary.
- Test a judicious selection of combinations of conditions. . . .
- For mission-critical software, or when the test budget allows, it is important to test beyond the required or designed limits to identify the points at which the system degrades or fails. [Green, 1987]

The majority of test cases in the representative set should be situations which the expert system typically handles. These test cases ideally cover a range of problem types and a range of problem complexity. [Tuthill, 1990] The test cases are documented in the appropriate test plans. Included in these test plans are the expected results of the test cases. An automated tool providing the user-friendly entry and maintenance of a repository of validation test cases is recommended. Such a tool is especially advantageous for large expert systems where many test cases are needed for comprehensive validation. An automated tool for analyzing the effectiveness of coverage for a set of validation test cases is also recommended. Additionally, the automated generation of suggestions for test cases based on coverage analysis is suggested for use in developing a comprehensive set of test cases for the validation process.

C. Acceptance Criteria

The establishment of acceptance criteria for rule-based expert systems is advocated in this book. Although the method of developing these acceptance criteria is beyond the scope of this book, certain aspects of expert systems and validation are enumerated here for inclusion in the acceptance criteria. Acceptance criteria should designate not only what is to be measured but also how it is to be measured, i.e., how the success or failure of the criteria is to be assessed. The acceptance criteria should describe the functional and performance criteria for the expert system in unambiguous, testable statements. Some measure of accuracy which the expert system must obtain should be established. For example, if human experts diagnose blood diseases with an accuracy of 70%, then the acceptance criteria should state that the expert system is expected to match the performance of the human expert(s) at 70%. The specific conditions under which the expert system is to operate should be defined in the acceptance criteria. For systems such as embedded expert systems, the functional requirements can be stated in relation to the inputs and outputs of the system which are clearly defined for these types of expert systems.

D. Testing Documentation

The need for testing documentation in the production of quality expert systems is gradually being recognized: "a solid test plan is another of those imperative necessities for a successful project." [Tuthill, 1990] The preparation of testing documentation is acknowledged as an effective means towards better understanding of the system under development. The preparation of testing documentation is also a necessity for many industrial and all military applications where the requirements of resources, time, and personnel needed for the testing process must be specified and precise acceptance standards must be stated.

Currently, the only testing documentation described for expert systems is the test plan, as seen in the Linear Model in Section III of this chapter. However, a Project V & V Plan is needed to define and delineate the verification and validation activities for the expert system under development. In order to provide systematic and comprehensive quality assurance for expert systems, the verification and validation, particular to the expert system under development, need to be specified in detail. The Project V & V Plan designates the exact verification and validation activities to be performed, the required inputs for these activities, and the outputs to be produced. The verification and validation activities for each phase in the process model are described. The Project V & V Plan also includes an overall schedule of all verification and validation activities, i.e, milestones, completion dates, the resources needed to complete these activities, the tools and techniques to be used to accomplish each of the specified activities, and the personnel responsible for performing the activities. The concept of the Project V & V Plan is derived from the testing documentation of conventional software, as described in Sections III and V of Chapter 3. The preparation of such documentation is necessary if a well organized, unambiguous, and methodical approach to verification and validation is to be established for expert systems.

Additionally, the test plan, presented in the Linear Model, describes the verification and validation testing activity for the development process as part of the design document. A better approach is the development of multiple test plans. A test plan for each type of testing activity delineated in the Project V & V Plan needs to be developed. Having separate test plans reflects the specific objectives for the different types of testing activities, separates the test data according to its testing activity, and, generally, reinforces the distinctiveness of the different types of verification and validation testing activities. A test plan for validation during development, a test plan for validation for customer acceptance, and a test plan for revalidation should be produced. The test cases within these three test plans may overlap.

E. Revalidation

As stated earlier, revalidation occurs during the operational use of the expert system. Revalidation is applied after the implementation of any enhancements or modifications to the expert system. The purpose of revalidation is to ensure that the level of expertise and accuracy of the expert system is the same as or better than before the changes and that none of the required functionality or performance of the expert system (as stated in the requirements specification) has been lost or degraded. Revalidation is achieved through the application of a set of test cases which have successfully tested the expert system in previous validation.

In order to effectively and efficiently perform revalidation, test cases in the repository from validation during development and validation for customer acceptance are utilized. Along with this suite of test cases, the results from the previous validation testing are also maintained. The comparison of the results from revalidation with the results of previous validation testing provides a measure of the effect of the modification or enhancement. Tools for the automated application of the suite of test cases and for automated record-keeping and report-generating are advantageous because their use eliminates the time-consuming and tedious nature of manual revalidation. The test cases in the repository are possibly supplemented with new test cases which are needed to more thoroughly validate the modification or enhancement. Revalidation should be part of the Project V & V Plan.

F. Validation Tools

The use of automated tools in the validation of expert systems has been suggested in the recommendations presented above. However, no such tools currently exist for the validation of expert systems. The need for validation tools has been recognized in past research on expert systems. Expert system development, on the whole, lacks tools and "will continue to be an art rather than an engineering discipline until tools become available to make the system builder's life easier." [Price, 1990] Furthermore, the scarcity of experts, the limitations on the time of experts, and the expense of employing experts in the development and validation of expert systems are reasons for more effectively utilizing the experts' time through the use of automated tools. [Kang, 1990]

The automated tools suggested in previous research on the validation of expert systems, which are also recommended in this book, include the maintenance of a repository of test cases, the application of test cases in validation testing, the comparison of test results, and maintenance of the record-keeping tasks of revalidation. Tools for the measurement of the effectiveness of test cases in validating an expert system, for the generation of test case suggestions based on this effectiveness, and for the production of statistical analysis of validation results are presented in this book for more effective validation of expert systems. Such validation tools will become more essential as expert systems become larger, as applications for expert systems become more critical, and as expert systems are maintained for longer periods of time. The potential of automated tools in the validation of expert systems is further discussed in Chapter 7.

VI. PREVALENT QUESTIONS ABOUT VERIFICATION AND VALIDATION

Seven questions, which are addressed in most discussions on the verification and validation of expert systems, are seven problems enumerated by O'Keefe et al. [O'Keefe, 1987] These seven questions are:

1. What to validate
2. What to validate against
3. What to validate with
4. When to validate
5. How to control the cost of validation
6. How to control bias
7. How to cope with multiple results. [O'Keefe, 1987]

In the following discussion, the definitions and recommendations of this book are examined in light of these seven questions.

What to validate? Validation of expert systems involves the validation of implicit and explicit requirements. The implicit requirement for expert systems is the equivalency in performance to human experts. The comparison of the performance of the expert system with the performance of human expert(s) is the primary technique in the determination of equivalency. If historical test cases are utilized in validation, then the known results act as the performance of the human expert in this comparison. Validation of explicit requirements is the validation of the acceptance criteria in the requirements specification document. The demonstration of these requirements is necessary to ensure customer acceptance of the expert system. Whether validation is measuring the equivalency of performance to human expert(s) or demonstrating required functionality, validation is based on the quality of the final decisions of an expert system. The three forms of validation testing, which are validation during development, validation for customer acceptance, and revalidation, examine the quality of the final decision of the expert system.

What to validate against? Expert systems are validated against human experts or known results. On-going transactions, real-world scenarios, and generated cases utilize answers from human expert(s) in the validation process. Historical cases provide known results. The answers of human expert(s) or the known results are compared against the answers of the expert system.

What to validate with? Validation is ideally accomplished through a comprehensive set of test cases from on-going transactions, real-world scenarios, and historical cases. These test cases represent a variety of problems and levels of difficulty which the expert system encounters. The availability of such test cases is domain specific. For domains which do not provide a sufficient number of test cases, test cases are generated by the domain expert or customer. Such test cases represent typical situations which the expert system encounters. All these test cases are stored in a repository of test cases for validation testing. After the execution of a suite of test cases in the repository, the effectiveness of validation coverage of these test cases is measured, and suggestions for new test cases, which assist in more comprehensively validating the expert system, are generated automatically.

When to validate? Validation activities occur throughout the development process of expert systems. Preliminary validation activities include the production of a Project V & V Plan, production of a test plan for the three types of validation testing, definition of the acceptance criteria in the requirements specification document, and collection and/or generation of validation test cases. Actual validation testing occurs during development, for customer acceptance, and during operational use. Validation during development occurs after the development team has completed the verification testing of the expert system. The development team then performs validation testing of the implicit and explicit requirements. After validation during development is satisfactorily achieved, the customer begins validation testing. Validation for customer acceptance is performed in the operational environment, if the application is noncritical. Revalidation, the validation testing during operational use, occurs after any modification to the expert system.

How to control the cost of validation? The use of automation controls the cost of validation. Automation of the maintenance of a test case repository, execution of test cases, application of statistical measures in analyzing the recommendations of an expert system against the recommendations of human expert(s), record-keeping and report-generating of previous validation testing, analysis of the effectiveness of test cases in the validation process, and generation of test case suggestions for more complete validation of the expert system provides an effective means of controlling the cost of validation. Such automation allows for effective use of the expensive and limited time of the domain experts.

How to control bias? The automation of the comparison of results of human expert(s) with the results of the expert system eliminates the bias of human judgement. As discussed in Section I, the manual comparison of these results is often biased by the human comparator's prejudice or parochialism.

How to cope with multiple results? Determining a clear success or failure is problematic with multiple results. The use of statistical measures for comparison is one means of coping with multiple results. The statistical measures for comparison presented in this book are order compatibility and distance comparability which are described in Section II of Chapter 7. These measures provide different viewpoints by which the equivalency of multiple results can be measured.

The recommendations presented in the previous sections address the difficulties and problems of validating rule-based expert systems. Misinterpretation of validation, present in the current approach to validation, is lessened with the definition of validation and the demarcation of verification and validation. The method of test case selection is improved with the suggestion of selection criteria, the usage of coverage analysis, and the generation of test cases for improving coverage. Automation of the execution of test cases and the comparison of results eliminates the time-consuming and tedious tasks of validation and eliminates the subjectivity of the comparison process. The need for preliminary planning activities is addressed in the preparation of the Project V & V Plan and test plans. The use of statistical measures provides a means of managing the complex results produced by expert systems. The measures of order compatibility and distance comparability provide different assessments of the equivalency between the performance of the expert system and human expert(s) where multiple results and certainty factors are involved. The feasibility of these recommendations is addressed in the next chapter.

7

VALIDATION
PROTOTYPE — SAVES

7

VALIDATION
PROTOTYPE — SAVES

I. INTRODUCTION AND ASSUMPTIONS

SAVES (Suzanne and Abe's Validator for Expert Systems) is a demonstration prototype developed to illustrate the feasibility of recommendations presented in this book. The recommendations demonstrated in SAVES establish a comprehensive set of techniques and tools for the validation of rule-based expert systems. These techniques and tools include maintenance of a test case repository, execution of test cases, application of statistical measures in analyzing the recommendations of an expert system against the recommendations of human expert(s), record-keeping and report-generating of previous validation testing, analysis of the effectiveness of test cases in the validation process, and generation of test case suggestions for more complete validation of the expert system.

Ideally, a validation tool, like SAVES, is part of a larger development environment which includes a knowledge base, inference engine, user interface, knowledge acquisition tools, and static and dynamic verification tools. Such a development environment is depicted in Figure 7.1. However, presently, no such environment is available. Therefore, the demonstration prototype SAVES is being developed with the knowledge engineering tool M.1 providing the knowledge base, inference engine, and user interface. M.1 is a commercial, PC-based development tool available from Cimflex Teknowledge. M.1 supports backward chaining, rules, and certainty factors. M.1 also allows direct access to the contents of knowledge base.

The prototype SAVES is designed to validate rule-based expert systems. The focus of the coverage analysis within SAVES is the validation of the knowledge base. The coverage of the rules within a knowledge base by validation test cases is analyzed in SAVES. The use of commercially available knowledge development tools, such as the M.1, allows the knowledge engineer to concentrate on validating the knowledge instead of the inference engine and user interface which are supplied by the knowledge development tool.

Classification and construction problems are the two major types of tasks solved by expert systems currently. Classification problems are tasks which have an enumerable set of solutions. Monitoring, prediction, and diagnosis are examples of classification problems. Construction problems build solutions from a set of components; the possible solutions are not enumerable. Design, planning, and repair are examples of construction problems.

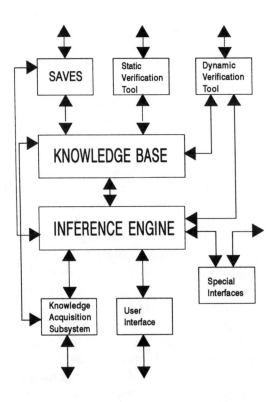

FIGURE 7.1.   Total Development Environment

SAVES is designed to validate rule-based expert systems developed to solve classification problems. The recommendations of this book focus on classification problems because they are "the most common (and fruitful) area of use for expert system shells." [Wilson, 1990] Additionally, classification problems are more suitable for automating the analysis of validation coverage. Determining if all possible solutions have been validated is viable only if all possible solutions are delineated.

In the description and evaluation of SAVES, the knowledge base utilized with M.1 is for the wine selection problem. In the wine selection problem, the elements of a menu and elements of wine preference including preferred body, preferred color, and preferred sweetness of wine are the basis for the selection of an appropriate wine. Using these elements as inputs, the expert system produces recommendations about the wine(s) suitable for the menu and preference elements. The recommendations of the expert system are composed of suggested wine(s) and their associated certainty factor(s). The contents of the knowledge base for the wine selection problem used in the description and evaluation of SAVES are as follows:

```
kb-1: goal = wine
rule1:
    if has-sauce=yes and
        sauce=spicy
    then best-body=full.
rule2:
    if tastiness=delicate
    then best-body=light cf 80.
rule3:
    if tastiness=average
    then best-body=light cf 30 and
        best-body=medium cf 60 and
        best-body=full cf 30.
rule4:
    if tastiness=strong
    then best-body=medium cf 40 and
        best-body=full cf 80.
rule5:
    if has-sauce=yes and
```

sauce=cream
then best-body=medium cf 40 and
best-body=full cf 60.

rule6:
if main-component=meat and
has-veal=no
then best-color=red cf 90.

rule7:
if main-component=poultry and
has-turkey=no
then best-color=white cf 90 and
best-color=red cf 30.

rule8:
if main-component=fish
then best-color=white.

rule9:
if not main-component=fish and
has-sauce=yes and
sauce=tomato
then best-color=red.

rule10:
if main-component=poultry and
has-turkey=yes
then best-color=red cf 80 and
best-color=white cf 50.

rule11:
if main-component is unknown and
has-sauce=yes and
sauce=cream
then best-color=white cf 40.

rule12:
if has-sauce=yes and
sauce=sweet
then best-sweetness=sweet cf 90 and
best-sweetness=medium cf 40.

rule13:
if has-sauce=yes and
sauce=spicy
then feature=spiciness.

rule14:

    if best-body=light
    then recommended-body=light.
rule15:
    if best-body=medium
    then recommended-body=medium.
rule16:
    if best-body=full
    then recommended-body=full.
rule17:
    if preferred-body=light and
        best-body=light
    then recommended-body=light cf 20.
rule18:
    if preferred-body=medium and
        best-body=medium
    then recommended-body=medium cf 20.
rule19:
    if preferred-body=full and
         best-body=full
    then recommended-body=full cf 20.
rule20:
    if preferred-body=light and
        best-body=full
    then recommended-body=medium.
rule21:
    if preferred-body=full and
        best-body=light
    then recommended-body=medium.
rule22:
    if preferred-body=light and
        best-body is unknown
    then recommended-body=light.
rule23:
    if preferred-body=medium and
        best-body is unknown
    then recommended-body=medium.
rule24:
    if preferred-body=full and
        best-body is unknown
    then recommended-body=full.

rule25:
   if best-body is unknown
   then recommended-body=medium.
rule26:
   if best-color=red
   then recommended-color=red.
rule27:
   if best-color=white
   then recommended-color=white.
rule28:
   if preferred-color=red and
       best-color=red
   then recommended-color=red cf 20.
rule29:
   if preferred-color=white and
       best-color=white
   then recommended-color=white cf 20.
rule30:
   if preferred-color=red and
       best-color is unknown
   then recommended-color=red.
rule31:
    if preferred-color=white and
        best-color is unknown
   then recommended-color=white.
rule32:
   if preferred-color is unknown
   then recommended-color=red cf 50 and
       recommended-color=white cf 50.
rule33:
   if best-sweetness=dry
   then recommended-sweetness=dry.
rule34:
   if best-sweetness=medium
   then recommended-sweetness=medium.
rule35:
   if best-sweetness=sweet
   then recommended-sweetness=sweet.
rule36:
   if best-sweetness is unknown and

    preferred-sweetness is unknown
  then recommended-sweetness=medium.
rule37:
    if best-sweetness=dry and
        preferred-sweetness=dry
    then recommended-sweetness=dry cf 20.
rule38:
    if best-sweetness=medium and
        preferred-sweetness=medium
    then recommended-sweetness=medium cf 20.
rule39:
    if best-sweetness=sweet and
        preferred-sweetness=sweet
    then recommended-sweetness=sweet cf 20.
rule40:
    if preferred-sweetness=dry and
        best-sweetness is unknown
    then recommended-sweetness=dry.
rule41:
    if preferred-sweetness=medium and
        best-sweetness is unknown
    then recommended-sweetness=medium.
rule42:
    if preferred-sweetness=sweet and
        best-sweetness is unknown
    then recommended-sweetness=sweet.
rule43:
    if preferred-sweetness=dry and
        best-sweetness=sweet
    then recommended-sweetness=medium.
rule44:
    if preferred-sweetness=sweet and
        best-sweetness=dry
    then recommended-sweetness=medium.
rule45:
    if recommended-color=red and
        recommended-body=medium and
        recommended-sweetness=medium or
        recommended-sweetness=sweet
    then wine=gamay.

rule46:

    if recommended-color=white and
       recommended-body=light and
       recommended-sweetness=dry
    then wine=chablis.

rule47:

    if recommended-color=white and
       recommended-body=medium and
       recommended-sweetness=dry
    then wine=sauvignon-blanc.

rule48:

    if recommended-color=white and
       recommended-body=medium or
       recommended-body=full and
       recommended-sweetness=dry or
       recommended-sweetness=medium
    then wine=chardonnay.

rule49:

    if recommended-color=white and
       recommended-body=light and
       recommended-sweetness=dry or
       recommended-sweetness=medium
    then wine=soave.

rule50:

    if recommended-color=white and
       recommended-body=light or
       recommended-body=medium and
       recommended-sweetness=medium or
       recommended-sweetness=sweet
    then wine=riesling.

rule51:

    if recommended-color=white and
       recommended-body=full and
       feature=spiciness
    then wine=gewuerztraminer.

rule52:

    if recommended-color=white and
       recommended-body=light and
       recommended-sweetness=medium or
       recommended-sweetness=sweet

then wine=chenin-blanc.
rule53:
    if recommended-color=red and
        recommended-body=light
    then wine=valpolicella.
rule54:
    if recommended-color=red and
        recommended-sweetness=dry or
        recommended-sweetness=medium
    then wine=cabernet-sauvignon and
        wine=zinfandel.
rule55:
    if recommended-color=red and
        recommended-body=medium and
        recommended-sweetness=medium
    then wine=pinot-noir.
rule56:
    if recommended-color=red and
        recommended-body=full
    then wine=burgundy.

The wine selection problem is utilized because of the size and complexity of the knowledge base and the availability of the knowledge base through M.1.

The concepts implemented in SAVES assist in the validation of rule-based expert systems throughout the development process and operational life of the systems. SAVES demonstrates that such a development tool is feasible and that the development of such a comprehensive set of techniques and tools for the validation of rule-based expert systems is advantageous.

II. COMPONENTS OF SAVES

The three main components of SAVES are the validation test manager, validation archiver, and validation analyzer and enhancer. The subsystem data flow diagram for SAVES containing these main components is shown in Figure 7.2. These components function together to automate the activities of validating a rule-based expert system.

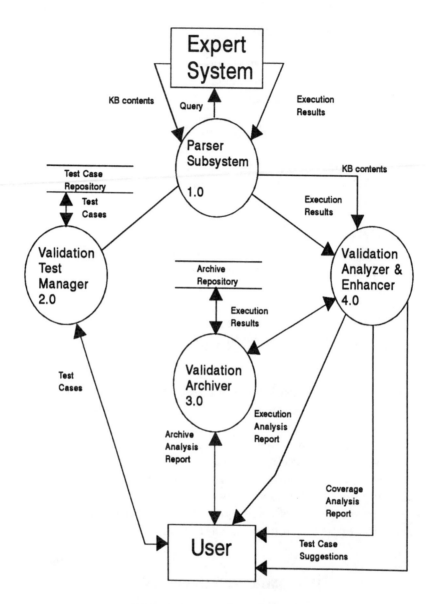

FIGURE 7.2. SAVES Subsystems

SAVES is a menu-driven system. Menus are used at every level in the system including much of the data entry during the creation of test cases. Figure 7.3 shows the main menu of SAVES and, thus, the functionality provided by these three main components. The

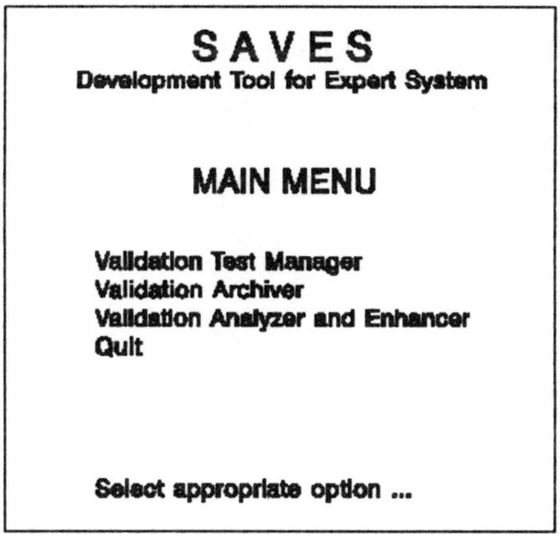

```
┌──────────────────────────────────────┐
│              S A V E S                │
│    Development Tool for Expert System │
│                                        │
│                                        │
│             MAIN MENU                  │
│                                        │
│                                        │
│      Validation Test Manager           │
│      Validation Archiver               │
│      Validation Analyzer and Enhancer  │
│      Quit                              │
│                                        │
│                                        │
│                                        │
│      Select appropriate option ...     │
│                                        │
└──────────────────────────────────────┘
```

FIGURE 7.3.   Main Menu for SAVES

functionality of each of the components within SAVES is described in the following sections.

A. Validation Test Manager

The validation test manager component oversees the creation, deletion, display, and execution of test cases. The functionality of each of these tasks is described in greater detail later in this section. The subsystem data flow diagram for the validation test manager component is shown in Figure 7.4.

In order to perform these functions, the validation test manager supports and facilitates the maintenance of a test case repository. The test cases used within SAVES are maintained in a representation independent of the rule-based expert system. The repository of test cases can be dynamically updated throughout the development process and operational use of the expert system using the test case creation facility within the validation test manager.

1. Creating Test Cases

The test case creation facility within this component allows the user to create test cases and maintain them in the repository of test cases within SAVES. A menu-driven system is used by the test case creation facility for the automated entry of test cases, i.e., not the automated generation of test cases. The automated entry of test

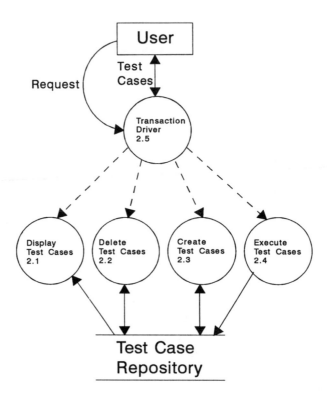

FIGURE 7.4.   Subsystem Data Flow Diagram for Validation Test Manager Component

cases is accomplished through an analysis of the knowledge base of the expert system being validated. This analysis is accomplished by the parser subsystem within SAVES which is described below. The subsystem data flow diagram for the test case creation facility is shown in Figure 7.5.

A parser subsystem within SAVES analyzes the knowledge base in order to determine the information necessary for the creation of a test case. In the parser subsystem, the antecedents and consequents of the knowledge base rules are analyzed for the delineation of test case variables and for the determination of all possible values for test case variables. For example, from the knowledge base of the wine selection problem, the parser subsystem ascertains that the test case variables are 'main-component', 'has-turkey', 'has-veal', 'has-sauce', 'sauce', 'tastiness', 'preferred-color', 'preferred-body', and 'preferred-sweetness'. The parser subsystem also determines all possible values for each test case

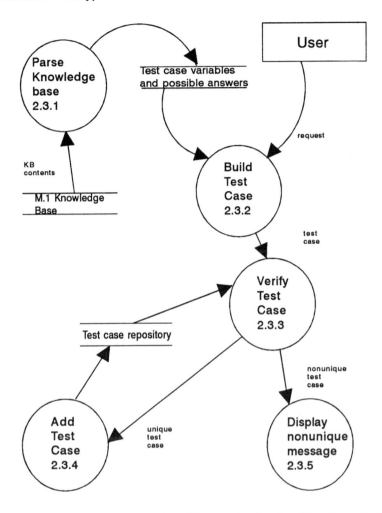

FIGURE 7.5. Subsystem Data Flow Diagram for Creating Test Cases Facility with Validation Test Manager Component.

variable; for example, the test case variable 'preferred-sweetness' has the possible values of 'sweet', 'medium', 'dry', and 'unknown'.

The parser subsystem, unlike most components within SAVES, is knowledge base specific. The syntax and semantics of the knowledge base being used, i.e., M.1 are known and prescribe the analysis of the knowledge base. The following segment of code from the parsing subsystem within the test case creation facility illustrates the need for understanding of the syntax and semantics of the M.1 knowledge base in order to derive the test case variables and their possible values:

```
assign(infilekb, kbfilename);
reset(infilekb);
repeat
 read(infilekb,ch);
 if ch = 'r' then
 begin
    {at the beginning of a rule}
    readln(infilekb);
    repeat
       read(infilekb, ch);
    until ch = 'i';
    repeat
       read(infilekb, ch);
    until ch = ' ';
       repeat
       read(infilekb, ch);
    until ch <> ' ';
    repeat
     more := false;
     I := 1;
     while (ch <> '=') and (ch <> ' ') do
     begin
      varword[I] := ch;
      I := I + 1;
      read(infilekb, ch);
     end; {while}
     for I := I to maxword do
         varword[I] := ' ';
     if (varword[1]='n') and (varword[2] = 'o')
         and (varword[3] = 't')
         and (varword[4] = ' ') then
     begin
      I := 1;
      read(infilekb,ch);
      while (ch <> '=') and (ch <> ' ') do
      begin
         varword[I] := ch;
         I := I + 1;
         read(infilekb, ch);
      end; {while}
```

```
      for I := I to maxword do
            varword[I] := ' ';
end; {if varword = not}
if ch = '=' then
begin
 {premise in format variable = value}
 I := 1;
 read(infilekb, ch);
 repeat
    valword[I] := ch;
    I := I + 1;
    read(infilekb, ch);
 until (ch = ' ') or (eoln(infilekb));
 if ch <> ' ' then
 begin
    valword[I] := ch;
    I := I + 1;
    end;
 for I := I to maxword do
     valword[I] := ' ';
end {if ch = '=' then}
else
begin
 {premise in format variable is value}
 read(infilekb, ch, ch);{read off is}
 read(infilekb, ch); {read off blank}
 I := 1;
 read(infilekb, ch);
 while (ch <> ' ') and
         (not eoln(infilekb)) do
 begin
    valword[I] := ch;
    I := I + 1;
    read(infilekb,ch);
 end; {while}
 if ch <> ' ' then
 begin
    valword[I] := ch;
    I := I + 1;
    end;
```

```
   for I := I to maxword do
       valword[I] := ' ';
   end; {else}
   CHECK_FOR_HAS (varword, hasflag, hasword);
   ADD_TO_LIST (varword, valword, false,
                       hasflag, hasword,
                       kbvariables, numvariables);
   if ch = ' ' then
   begin
     read(infilekb,ch);
     if (ch = 'a') or (ch = 'o') then
     begin
       more := true;
       readln(infilekb);
       repeat
        read(infilekb,ch);
     until (ch <> ' ');
     end; {if ch = a or o}
   end; {if ch = blank}
until more = false;

{handle conclusion of rule}
   readln(infilekb);
repeat
   read(infilekb, ch);
until ch = 't';
repeat
   read(infilekb, ch);
until ch = ' ';
repeat
   read(infilekb, ch);
until ch <> ' ';
repeat
   more := false;
   extraconclus := false;
   I := 1;
   while (ch <> '=') do
   begin
    varword[I] := ch;
    I := I + 1;
```

```
   read(infilekb,ch);
end;
for I := I to maxword do
     varword[I] := ' ';
{read value to variable}
I := 1;
read(infilekb,ch);
while (ch <> ' ') and (ch <> '.') do
begin
 valword[I] := ch;
 I := I + 1;
 read(infilekb, ch);
end;
if ch = ' ' then
begin
 read(infilekb,ch);
 if (ch = 'c') then
 begin
    {read certainty factor of value}
    valword[I] := ' ';
    I := I + 1;
    valword[I] := ch;
    I := I + 1;
    read(infilekb,ch);
    valword[I] := ch;
    I := I + 1;
    read(infilekb,ch);
    valword[I] := ch;
    I := I + 1;
    read(infilekb,ch);
    repeat
     valword[I] := ch;
     I := I + 1;
     read(infilekb, ch);
  until (ch = ' ') or (ch = '.');
  end {if ch = c}
  else if (ch = 'a') or (ch = 'o') then
  begin
     more := true;
     extraconclus := true;
```

```
        readln(infilekb);
        repeat
            read(infilekb, extrach);
        until extrach <> ' ';
      end; {elseif ch = a or o}
    end; {if ch = blank}
    for I := I to maxword do
        valword[I] := ' ';
    ADD_TO_LIST(varword, valword, true, false,
                    hasword, kbvariables,
                    numvariables);
if ch = '.' then
    more := false
else if extraconclus = false then
begin
    read(infilekb,ch);
    if (ch = 'a') or (ch = 'o') then
    begin
    more := true;
    readln(infilekb);
    repeat
        read(infilekb, ch);
    until ch <> ' ';
    end; {if ch = a or o}
end; {else}
if extraconclus = true then
    ch := extrach;
until more = false;
end {if ch = r}
else if ch = 'k' then
begin
{handle the kbXX lines}
valword := blank25;
varword := blank25;
repeat
    read(infilekb, ch);
until (ch = ' ');
repeat
    read(infilekb, ch);
until (ch <> ' ');
```

```
I := 1;
while (ch <> ' ') and (ch <> '=') do
begin
   varword[I] := ch;
   I := I + 1;
   read (infilekb, ch);
end;
for I := I to maxword do
      varword[I] := ' ';
if ch = '=' then
begin
   repeat
    read (infilekb, ch);
   until (ch <> ' ');
   I := 1;
   while (ch <> ' ') and not eoln (infilekb)
   do begin
    valword[I] := ch;
    I := I + 1;
    read (infilekb, ch);
   end;
   if ch <> ' ' then
   begin
    valword[I] := ch;
    I := I + 1;
   end;
   for I := I to maxword do
       valword[I] := ' ';
end
else if ch = ' ' then
begin
   repeat
    read (infilekb,ch);
   until (ch = '=');
    repeat
    read(infilekb, ch);
   until (ch <> ' ');
   I := 1;
   while (ch <> ' ') and not eoln (infilekb)
   do begin
```

```
   valword[I] := ch;
    I := I + 1;
    read (infilekb, ch);
   end;
   if ch <> ' ' then
   begin
    valword[I] := ch;
    I := I + 1;
   end;
   for I := I to maxword do
        valword[I] := ' ';
 end;
 if (varword[1] = 'g') and
    (varword[2] = 'o') and
    (varword[3] = 'a') and
    (varword[4] = 'l') then
 begin
    kbgoal := valword;
 end;
 end; {else}
 readln(infilekb);
until eof (infilekb);
close(infilekb);
FIX_VALUES(kbvariables, numvariables);
```

This segment of code also illustrates the complexity of parsing even a relatively small knowledge base.

Upon request, the test case creation facility queries for the inputs of a test case using a menu-driven format of variables and possible values. Figures 7.6 through 7.15 illustrate the series of menus provided when entering values for a test case using the automated entry facility. As seen in these menus, all possible values for a variable are provided in a menu format. The desired value is then selected by simply moving the cursor onto that value in the menu and pressing the enter key.

After the entry of the necessary values of a test case, the test case is examined against the existing test cases in the repository for its uniqueness. Duplicated test cases are not permissible in the repository. An appropriate message is displayed to the user on the terminal screen in the event of the entry of a test case which

S A V E S
Development Tool for Expert System

MENU FOR VALIDATION TEST MANAGER

Creating Test Cases
Displaying Test Cases
Deleting Test Cases
Return to Main Menu

Select appropriate option ...

FIGURE 7.6. Menu for Validation Test Manager Component within SAVES

S A V E S
Development Tool for Expert System

CREATING TEST CASES
Automated Entry Mode

yes
no

Select the value for the variable
has-sauce

Select appropriate option ...

FIGURE 7.7. First Menu in Sequence for Creating Test Case

S A V E S
Development Tool for Expert System

CREATING TEST CASES
Automated Entry Mode

cream
spicy
sweet
tomato

Select the value for the variable
sauce

Select appropriate option ...

FIGURE 7.8. Second Menu in Sequence for Creating Test Case

S A V E S
Development Tool for Expert System

CREATING TEST CASES
Automated Entry Mode

average
delicate
strong

Select the value for the variable
tastiness

Select appropriate option ...

FIGURE 7.9. Third Menu in Sequence for Creating Test Case

S A V E S
Development Tool for Expert System

CREATING TEST CASES
Automated Entry Mode
fish
meat
poultry
unknown

Select the value for the variable
main-component

Select appropriate option ...

FIGURE 7.10. Fourth Menu in Sequence for Creating Test Case

S A V E S
Development Tool for Expert System

CREATING TEST CASES
Automated Entry Mode

yes
no

Select the value for the variable
has-veal

Select appropriate option ...

FIGURE 7.11. Fifth Menu in Sequence for Creating Test Case

S A V E S
Development Tool for Expert System

CREATING TEST CASES
Automated Entry Mode

yes
no

Select the value for the variable
has-turkey

Select appropriate option ...

FIGURE 7.12. Sixth Menu in Sequence for Creating Test Case

S A V E S
Development Tool for Expert System

CREATING TEST CASES
Automated Entry Mode

full
light
medium

Select the value for the variable
preferred-body

Select appropriate option ...

FIGURE 7.13. Seventh Menu in Sequence for Creating Test Case

```
┌─────────────────────────────┐  ┌─────────────────────────────┐
│                             │  │                             │
│         S A V E S           │  │         S A V E S           │
│ Development Tool for Expert  │  │ Development Tool for Expert  │
│           System            │  │           System            │
│                             │  │                             │
│     CREATING TEST CASES     │  │     CREATING TEST CASES     │
│     Automated Entry Mode    │  │     Automated Entry Mode    │
│                             │  │                             │
│  red                        │  │  dry                        │
│  white                      │  │  medium                     │
│  unknown                    │  │  sweet                      │
│                             │  │  unknown                    │
│  Select the value for the   │  │                             │
│  variable                   │  │  Select the value for the   │
│  preferred-color            │  │  variable                   │
│                             │  │  preferred-sweetness        │
│  Select appropriate option  │  │                             │
│  ...                        │  │  Select appropriate option  │
│                             │  │  ...                        │
│                             │  │                             │
└─────────────────────────────┘  └─────────────────────────────┘
```

FIGURE 7.14. Eighth Menu in Sequence for Creating Test Case

FIGURE 7.15. Ninth Menu in Sequence for Creating Test Case

already exists in the repository of test cases in SAVES.

For each unique test case, the human answers for the test case are queried for by the validation test manager. As with the entry of the other test case variables, the entry of possible solutions is also menu driven. The user is allowed to select from a list of all possible solutions for a test case. The list of possible solutions is determined from the analysis of the knowledge base by the parser subsystem within the validation test manager in the same manner that the test case variables and their possible answers are obtained. These possible solutions are derived from the conclusions in the knowledge base being validated. Answers from multiple experts are allowed. The certainty factors for possible solutions to the test case are also obtained.

2. Displaying Test Cases

The validation test manager within SAVES also provides facilities for displaying the test cases within the repository. Test cases are displayed to the screen or to the printer. The test cases are presented in an easy-to-read, tabular format. An example of the output displayed to the screen is shown in Figure 7.16.

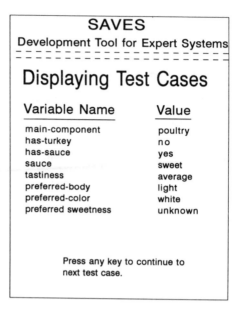

FIGURE 7.16.   Screen Output for Displaying Test Case Request

3. Deleting Test Cases

The deletion facility within the validation test manager allows all or part of the test cases to be removed from the test case repository within SAVES. To delete part of the contents of the repository, the test cases held with the repository are presented on the screen in a tabular format one at a time. At this time, the user of SAVES marks an individual test case for deletion. Multiple test cases can be marked for deletion. The deletion facility within the validation test manager provides security measures in case of accidental deletion of a test case.

4. Executing Test Cases

In the execution facility, either all or part of the test cases within the test case repository may be specified for inclusion in the test queue or execution suite, i.e., a suite of test cases to be executed by the expert system. The test cases to be included in the execution suite are selected by the user of SAVES in the same manner by which test cases are deleted from the test case repository, as described in the previous section.

The execution facility is designed to receive the execution suite, translate one test case at a time into the specified format,

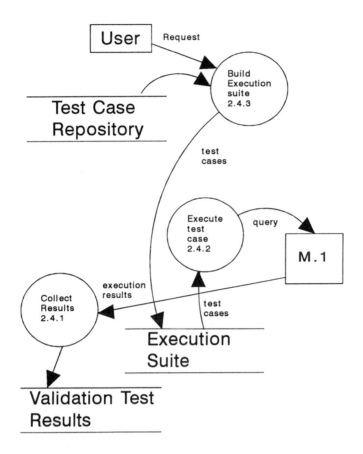

FIGURE 7.17. Subsystem Data Flow Diagram for Executing Test Cases Facility within Validation Test Manager Component

activate M.1, and receive and maintain execution information produced by M.1. The subsystem data flow diagram for the execution facility within the validation test manager is shown in Figure 7.17.

The development tool M.1 is adaptable to the requirements of the validation test manager of SAVES. These requirements are to receive one test case in the format specified by M.1, execute the test case, and return the recommendations of the expert system and the rules which were examined and/or fired by the expert system. A recommendation of the expert system consists of the answer and its certainty factor. SAVES is able to handle multiple recommendations by the expert system. The following is an example of the execution results produced by M.1 for one test case:

M.1> trace on
M.1> go
Seeking wine.
Invoking rule-33:
    if recommended-color = red and
        recommended-body = medium and
        (recommended-sweetness = medium or
        recommended-sweetness = sweet)
    then wine = gamay.
Seeking recommended-color.
Invoking rule-21:
    if best-color = red
    then recommended-color = red.
Seeking best-color.
Invoking rule-6:
    if main-component = meat and
        has-veal = no
    then best-color = red cf 90.
Seeking main-component.
Using kb-10:
    question(main-component) =
    'Is the main component of the meal meat, fish or poultry?'.
Is the main component of the meal meat, fish or poultry?
>> unknown
Noting main-component is unknown.
rule-6 failed.
Invoking rule-7:
    if main-component = poultry and
        has-turkey = no
    then best-color = white cf 90 and
           best-color = red cf 30.
Already sought main-component.
rule-7 failed.
Invoking rule-8:
    if main-component = fish
    then best-color = white.
Already sought main-component.
rule-8 failed.

Invoking rule-9:
    if not main-component = fish and
       has-sauce = yes and
       sauce = tomato
    then best-color = red.
Already sought main-component.
Seeking has-sauce.
Using kb-3:
    question(has-sauce) = 'Does the meal have a sauce on it?'.
Does the meal have a sauce on it?
>> yes
Noting has-sauce = yes cf 100 because you said so.
Found has-sauce.
Seeking sauce.
Using kb-22:
    question(sauce) =
    'Is the sauce for the meal spicy, sweet, cream or tomato?'.
Is the sauce for the meal spicy, sweet, cream or tomato?
>> sweet
Noting sauce = sweet cf 100 because you said so.
Found sauce.
rule-9 failed.
Invoking rule-10:
    if main-component = poultry and
       has-turkey = yes
    then best-color = red cf 80 and
        best-color = white cf 50.
Already sought main-component.
rule-10 failed.
Invoking rule-11:
    if has-sauce = yes and
       sauce = cream
    then best-color = white cf 40 and
        best-color = red cf -90.
Already sought has-sauce.
Already sought sauce.
rule-11 failed.
Noting best-color is unknown.
rule-21 failed.
Invoking rule-22:

        if best-color = white
        then recommended-color = white.
Already sought best-color.
rule-22 failed.
Invoking rule-23:
        if best-color is unknown and
            preferred-color = red
        then recommended-color = red.
Already sought best-color.
Seeking preferred-color.
Using kb-16:
        question(preferred-color) = 'Do you generally prefer red or
white wines?'.
Do you generally prefer red or white wines?
>> red
Noting preferred-color = red cf 100 because you said so.
Found preferred-color.
Noting recommended-color = red cf 100 because rule-23.
rule-23 succeeded.
Found recommended-color.
Seeking recommended-body.
Invoking rule-14:
        if best-body = light
        then recommended-body = light.
Seeking best-body.
Invoking rule-1:
        if has-sauce = yes and
            sauce = spicy
        then best-body = full.
Already sought has-sauce.
Already sought sauce.
rule-1 failed.
Invoking rule-2:
        if tastiness = delicate
        then best-body = light cf 80.
Seeking tastiness.
Using kb-25:
        question(tastiness) =
        'Is the flavor of the meal delicate, average or strong?'.
Is the flavor of the meal delicate, average or strong?

>> average
Noting tastiness = average cf 100 because you said so.
Found tastiness.
rule-2 failed.
Invoking rule-3:
    if tastiness = average
    then best-body = light cf 30 and
            best-body = medium cf 60 and
            best-body = full cf 30.
Already sought tastiness.
Noting best-body = light cf 30 because rule-3.
Noting best-body = medium cf 60 because rule-3.
Noting best-body = full cf 30 because rule-3.
rule-3 succeeded.
Invoking rule-4:
    if tastiness = strong
    then best-body = medium cf 40 and
            best-body = full cf 80.
Already sought tastiness.
rule-4 failed.
Invoking rule-5:
    if has-sauce = yes and
        sauce = cream
    then best-body = medium cf 40 and
            best-body = full cf 60.
Already sought has-sauce.
Already sought sauce.
rule-5 failed.
Found best-body.
Noting recommended-body = light cf 30 because rule-14.
rule-14 succeeded.
Invoking rule-15:
    if best-body = medium
    then recommended-body = medium.
Already sought best-body.
Noting recommended-body = medium cf 60 because rule-15.
rule-15 succeeded.
Invoking rule-16:
    if best-body = full
    then recommended-body = full.

Already sought best-body.
Noting recommended-body = full cf 30 because rule-16.
rule-16 succeeded.
Invoking rule-17:
    if best-body is unknown and
        preferred-body = light
    then recommended-body = light.
Already sought best-body.
rule-17 failed.
Invoking rule-18:
    if best-body is unknown and
        preferred-body = medium
    then recommended-body = medium.
Already sought best-body.
rule-18 failed.
Invoking rule-19:
    if best-body is unknown and
        preferred-body = full
    then recommended-body = full.
Already sought best-body.
rule-19 failed.
Invoking rule-20:
    if best-body is unknown and
        preferred-body is unknown
    then recommended-body = medium.
Already sought best-body.
rule-20 failed.
Found recommended-body.
Seeking recommended-sweetness.
Invoking rule-26:
    if best-sweetness = dry
    then recommended-sweetness = dry.
Seeking best-sweetness.
Invoking rule-12:
    if has-sauce = yes and
        sauce = sweet
    then best-sweetness = sweet cf 90 and
           best-sweetness = medium cf 40.
Already sought has-sauce.
Already sought sauce.

Noting best-sweetness = sweet cf 90 because rule-12.
Noting best-sweetness = medium cf 40 because rule-12.
rule-12 succeeded.
Found best-sweetness.
rule-26 failed.
Invoking rule-27:
     if best-sweetness = medium
     then recommended-sweetness = medium.
Already sought best-sweetness.
Noting recommended-sweetness = medium cf 40 because rule-27.
rule-27 succeeded.
Invoking rule-28:
     if best-sweetness = sweet
     then recommended-sweetness = sweet.
Already sought best-sweetness.
Noting recommended-sweetness = sweet cf 90 because rule-28.
rule-28 succeeded.
Invoking rule-29:
     if best-sweetness is unknown and
          preferred-sweetness = dry
     then recommended-sweetness = dry.
Already sought best-sweetness.
rule-29 failed.
Invoking rule-30:
     if best-sweetness is unknown and
          preferred-sweetness = medium
     then recommended-sweetness = medium.
Already sought best-sweetness.
rule-30 failed.
Invoking rule-31:
     if best-sweetness is unknown and
          preferred-sweetness = sweet
     then recommended-sweetness = sweet.
Already sought best-sweetness.
rule-31 failed.
Invoking rule-32:
     if best-sweetness is unknown and
          preferred-sweetness is unknown
     then recommended-sweetness = medium.
Already sought best-sweetness.

rule-32 failed.

Found recommended-sweetness.

Noting wine = gamay cf 40 because rule-33.

Already sought recommended-sweetness.

Noting wine = gamay cf 60 because rule-33.

rule-33 succeeded.

Invoking rule-34:

    if recommended-color = white and
        recommended-body = light and
        recommended-sweetness = dry
    then wine = chablis.

Already sought recommended-color.

rule-34 failed.

Invoking rule-35:

    if recommended-color = white and
        recommended-body = medium and
        recommended-sweetness = dry
    then wine = sauvignon-blanc.

Already sought recommended-color.

rule-35 failed.

Invoking rule-36:

    if recommended-color = white and
        (recommended-body = medium or
        recommended-body = full) and
        (recommended-sweetness = dry or
        recommended-sweetness = medium)
    then wine = chardonnay.

Already sought recommended-color.

rule-36 failed.

Invoking rule-37:

    if recommended-color = white and
        recommended-body = light and
        (recommended-sweetness = dry or
        recommended-sweetness = medium)
    then wine = soave.

Already sought recommended-color.

rule-37 failed.

Invoking rule-38:

    if recommended-color = white and
        (recommended-body = light or

recommended-body = medium) and
(recommended-sweetness = medium or
recommended-sweetness = sweet)
then wine = riesling.
Already sought recommended-color.
rule-38 failed.
Invoking rule-39:
if recommended-color = white and
recommended-body = full and
feature = spiciness
then wine = gewuerztraminer.
Already sought recommended-color.
rule-39 failed.
Invoking rule-40:
if recommended-color = white and
recommended-body = light and
(recommended-sweetness = medium or
recommended-sweetness = sweet)
then wine = chenin-blanc.
Already sought recommended-color.
rule-40 failed.
Invoking rule-41:
if recommended-color = red and
recommended-body = light
then wine = valpolicella.
Already sought recommended-color.
Already sought recommended-body.
Noting wine = valpolicella cf 30 because rule-41.
rule-41 succeeded.
Invoking rule-42:
if recommended-color = red and
(recommended-sweetness = dry or
recommended-sweetness = medium)
then wine = cabernet-sauvignon and
wine = zinfandel.
Already sought recommended-color.
Already sought recommended-sweetness.
Already sought recommended-sweetness.
Noting wine = cabernet-sauvignon cf 40 because rule-42.
Noting wine = zinfandel cf 40 because rule-42.

rule-42 succeeded.
Invoking rule-43:
    if recommended-color = red and
        recommended-body = medium and
        recommended-sweetness = medium
    then wine = pinot-noir.
Already sought recommended-color.
Already sought recommended-body.
Already sought recommended-sweetness.
Noting wine = pinot-noir cf 40 because rule-43.
rule-43 succeeded.
Invoking rule-44:
    if recommended-color = red and
        recommended-body = full
    then wine = burgundy.
Already sought recommended-color.
Already sought recommended-body.
Noting wine = burgundy cf 30 because rule-44.
rule-44 succeeded.
Found wine.
    wine = gamay (76%) because rule-33.
    wine = cabernet-sauvignon (40%) because rule-42.
    wine = zinfandel (40%) because rule-42.
    wine = pinot-noir (40%) because rule-43.
    wine = valpolicella (30%) because rule-41.
    wine = burgundy (30%) because rule-44.
M.1> quit

This extensive data is analyzed by the SAVES in order to obtain information for the validation analyzer and enhancer component. The information obtained from this data includes not only the recommendations by the expert system but also the rules which were fired for this test case.

The section of code within test case execution facility which analyzes this output from M.1 is similar to the parser subsystem within SAVES in that it is M.1 dependent. In order to derive the necessary information from the extensive data created by the M.1 from the execution of one test case, the test case execution facility requires understanding of the format of the output produced by M.1. The following segment of code shows the M.1 dependent code to accomplish this analysis:

```
assign (infileES, ESfilename); {FILE CREATED BY M.1}
reset (infileES);
foundgoal := false;
numESresults := 0;
numrules := 0;
while not eof(infileES) do
begin
 for X := 1 to 80 do
     temp80[X] := ' ';
 X := 1;
 while not eoln(infileES) do
 begin
   read (infileES, temp80[X]);
   X := X + 1;
 end;
 readln(infileES);
 X := 0;
 repeat
   X := X + 1;
 until (temp80[X] <> ' ') or (X = 80);
 if foundgoal = false then
 begin
 if (temp80[X] = 'S') and
     (temp80[X + 1] = 'e') and
     (temp80[X + 2] = 'e') and
     (temp80[X + 3] = 'k') and
     (temp80[X + 4] = 'i') then
 begin
   foundgoal := true;
   repeat
    X := X + 1;
   until (temp80[X] = ' ');
   repeat
    X := X + 1;
   until (temp80[X] <> ' ');
   I := 1;
   repeat
    goal[I] := temp80[X];
```

```
    I := I + 1;
    X := X + 1;
    until (temp80[X] = '.') or
          (temp80[X] = ' ');
    for I := I to maxword do
        goal[I] := ' ';
  end; {if temp80 = seeking name of goal}
end; {if foundgoal = false}

if (temp80[X] = 'r') and
    (temp80[X + 1] = 'u') and
    (temp80[X + 2] = 'l') and
    (temp80[X + 3] = 'e') and
    (temp80[X + 4] = '-') then
begin
  for I := 1 to 3 do
      temp3[I] := ' ';
repeat
  X := X + 1;
until temp80[X] = '-';
X := X + 1;
I := 1;
repeat
  temp3[I] := temp80[X];
  I := I + 1;
  X := X + 1;
until (temp80[X] = ' ') or (I = 3);
for I := I to 3 do
      temp3[I] := ' ';
repeat
  X := X + 1;
until (temp80[X] <> ' ');
if (temp80[X] = 's') and
    (temp80[X + 1] = 'u') and
    (temp80[X + 2] = 'c') and
    (temp80[X + 3] = 'c') then
begin
  numrules := numrules + 1;
  assign(outfile, 'saveinfo.val');
  rewrite(outfile);
```

```
      writeln(outfile, temp3);
      close(outfile);
      reset(outfile);
      readln(outfile, rulenum);
      close(outfile);
      thecase.rulesexecuted[rulenum] := true;
      erase(outfile);
  end; {if}
  end;
  if (temp80[X] = goal[1]) and
      (temp80[X + 1] = goal[2]) and
      (temp80[X + 2] = goal[3]) and
      (temp80[X + 3] = goal[4]) then
  begin
    repeat
    X := X + 1;
  until (temp80[X] = ' ');
  repeat
    X := X + 1;
  until (temp80[X] <> ' ');
  repeat
    X := X + 1;
  until (temp80[X] = ' ');
  repeat
    X := X + 1;
  until (temp80[x] <> ' ');
  I := 1;
  repeat
    temp60[I] := temp80[X];
    I := I + 1;
    X := X + 1;
  until (temp80[X] = ' ');
  temp60[I] := ' ';
  I := I + 1;
  temp60[I] := ' ';
  I := I + 1;
  repeat
    X := X + 1;
  until (temp80[X] = '(') or
        (temp80[X] = '.');
```

```
if (temp80[X] = '(') then
begin
   numESresults := numESresults + 1;
   X := X + 1;
   repeat
    temp60[I] := temp80[X];
    I := I + 1;
    X := X + 1;
   until (temp80[X] = '%');
   for I := I to 80 do
        temp60[I] := ' ';
   thecase.ESresults[numESresults] := temp60;
 end; {if temp80 = (}
 end; {if temp80 = goal found}
end; {while not EOF}
close (infileES);
```

The information about the results of the validation testing per-formed by the validation test manager is available in the validation analyzer and enhancer component of SAVES. These results are described in Section C of this chapter.

B. Validation Archiver

Summary information about previous validation testing per-formed within SAVES is provided by the validation archiver com-ponent of SAVES. This information assists in analyzing the success-fulness and effectiveness of current validation testing in light of previous validation testing. The information includes summary execution and coverage information about test cases in the execu-tion suites of previous validation testing. Information about the status of the knowledge base and the test case repository at the time of the previous validation testing is also provided by the validation archiver. An example of the report produced by the validation archiver is shown in Table 7.1. This report is displayable to the screen or to the printer.

C. Validation Analyzer and Enhancer

Information on the results of validation testing achieved in the validation test manager and on the effectiveness of execution suite in validating the knowledge base is supplied by the validation

Table 7.1
Archive Report

VALIDATION ARCHIVE

Validation Report Number 1

SUMMARY INFORMATION

Time/Date Stamp: 11:20:07 10/19/90
Number of Test Cases in Execution Suite: 4
Number of Test Cases in Repository: 6
Number of Knowledge Base Rules: 56
Overall Coverage Index: 0.55

EXECUTION SUITE INFORMATION

Test Case Order Number	Compat.	Distance Compar.	Coverage Index	Effectivenes Index
1	1.00	82.61	0.13	0.04
2	1.00	289.94	0.34	0.20
3	1.00	77.56	0.20	0.05
4	1.00	156.12	0.14	0.09

analyzer and enhancer component within SAVES. Additionally, based on the information about the effectiveness of the test case coverage, test case suggestions are generated by the validation analyzer and enhancer. The validation analyzer and enhancer component within SAVES is comprised of three subsystems. These three subsystems, which are execution analysis, coverage analysis, and test case suggestions, are described in the following sections. The subsystem data flow diagram for the validation analyzer and enhancer component is shown in Figure 7.18.

1. Execution Analysis

A report on the results from validation testing achieved by the execution suite of test cases is generated by the execution analysis subsystem within the validation analyzer and enhancer component of SAVES. The comparison of the answers of human expert(s) with the answers produced by the expert system is shown in this report.

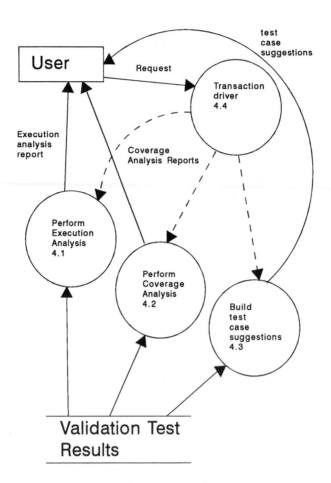

FIGURE 7.18. Subsystem Data Flow Diagram for Validation Analyzer and Enhancer Component

This comparison is enhanced with the use of statistical measures.

The application of statistical measures in validation has been proposed for both conventional software [Cho, 1987; Mills, 1987; Musa, 1989] and expert systems. [O'Keefe, 1987; Parsaye, 1988] In conventional software validation, statistical measures such as mean time to failure are utilized to evaluate the reliability of software. In expert system development, statistical measures have been suggested for comparing the answers of the expert system with the answers of human expert(s). Such statistical measures provide an objective assessment of the equivalency between the two sets of answers. These statistical measures eliminate the problematic evaluation of multiple answers and certainty factors.

In the execution analysis subsystem, the statistical measures of order compatibility and distance comparability are utilized. The use of order compatibility and distance comparability is suggested for the validation of expert systems by Parsaye. [Parsaye, 1988] These two statistical measures provide distinct points of view for assessing the equivalency between the recommendations of the expert system and the human expert(s) during validation.

In order compatibility, the order of recommendation in both sets of answers is the basis for assessing the equivalency between the performance of the expert system and the performance of human expert(s). The certainty factor for each answer indicates the order of recommendation. The higher certainty factor equates to the higher recommendation of that answer. Order compatibility does not measure that the certainty factors in two sets of answers match numerically but that the order of the answers (based on the certainty factors) match. For example, the expert system may respond with the following answers:

chardonnay: 85%
sauvignon-blanc: 80%
zinfandel: 40%
chablis: 30%

One human expert may respond to the same problem with answers such as:

chardonnay: 90%
sauvignon-blanc: 80%
zinfandel: 60%
chablis: 30%

Although the certainty factors in these two sets of answers do not match exactly, the order of the answers based on the certainty factors is in agreement. Therefore, these two sets of answers are order compatible. Sets of answers are order compatible if the order, from highest to lowest, of the certainty factors for the answers is the same in both sets of answers.

Complete order compatibility is not always necessary in expert system validation. Partial order compatibility may be sufficient for assessment. Sets of answers are partially order compatible with order N if the first N answers are in the same order, i.e., the certainty factors of the first N answers are in the same order from highest to lowest. For example, the expert system may respond with the following answers:

chardonnay: 85%
sauvignon-blanc: 80%
chablis: 40%
zinfandel: 20%

One human expert may respond to the same problem with answers such as:

chardonnay: 90%
sauvignon-blanc: 80%
zinfandel: 60%
chablis: 30%

These two sets of answers are partially order compatible with order 2 because the first two answers of each set are recommended in the same order, as indicated by the certainty factors of these two answers.

The concept of order compatibility is quantifiable by using a statistical measure such as Kendall's tau. [Snedecor, 1989] Kendall's tau provides a measure of the degree of concordance between two sets of answers, with +1 being complete concordance and -1 being complete disagreement. The partial order compatibility example, shown above, is used to describe the computation of Kendall's tau. The answers are ranked according to the order of recommendation reflected by the certainty factor. The rankings are rearranged so that the rankings of the human expert are in order from 1,2,3,...,n. Let chardonnay be numbered wine 1, sauvignon-blanc 2, chablis 3, and zinfandel 4. The rankings for this example are:

Wine Number	1	2	4	3
Human Expert	1	2	3	4
Expert System	1	2	4	3

Using the ranks of the expert system, since the ranks of the human expert are in order, the number of smaller ranks to the right of each rank are counted. These counts are 0, 0, and 1 for the first three ranks since no count can be taken for the rightmost rank. These counts are then added so that the total is 1; thus, Q = 1. Kendall's tau is:

$$tau = 1 - 4Q/[n(n - 1)]$$

For this example, tau = 1 - 4Q/[n(n - 1)] = 1 - 4/12 = 8/12 = 0.667.

In some situations, order compatibility may not represent a complete assessment of the equivalence between two sets of answers. For example, the expert system may respond with the following answers:

chardonnay: 95%
sauvignon-blanc: 85%
chablis: 60%
zinfandel: 50%

One human expert may respond to the same problem with answers such as:

chardonnay: 50%
sauvignon-blanc: 30%
chablis: 20%
zinfandel: 10%

These two sets of answers are order compatible, but the equivalency of these recommendations may be less than the measure of order compatibility can reflect.

Distance comparability provides an alternative measure of equivalence between the performance of the expert system and human expert(s). Distance comparability is utilized when the sets of numbers are ratio-scaled, i.e., the answer with a certainty factor of 60 is recommended twice as much as the answer with a certainty factor of 30. Distance comparability measures the distance between the two sets of answers as measured on all the certainty factors. A lower measure of distance comparability equates to closer equivalency between the answers of the expert system and the answers of the human expert. Where A is the expert system answers, B the human expert answers, and $CF(A_j)$ is the sum of the j certainty factors of the expert system, distance comparability is

$$[CF(A_j)^2 - CF(B_j)^2]^{1/2}$$

For the example above, the distance comparability is $[(95 + 85 + 60 + 50)^2] - [(50 + 30 + 20 + 10)^2]^{1/2} = [84100 - 12100]^{1/2} = 72000^{1/2} = 268.328$.

The following segment of code from the execution analysis subsystem shows the processing performed in order to calculate order compatibility using Kendall's tau:

```
FILL_IN_LISTS(ESres, humans);
SORT_ONE_ON_CF (humans);
for I := 1 to humans.howmanyhumans do
     DO_RANK(humans.onehuman[I]);
SORT_ONE_ES_ON_CF (ESres);
DO_RANK (ESres);
for I := 1 to humans.onehuman[1].numvalues do
 for J := 1 to ESres.numvalues do
 if ESres.answers[J].value =
     humans.onehuman[1].answers[I].value
 then
     temp.answers[I] := ESres.answers[J];

match := 0;
I := 1;
quit := false;
while (I <= ESres.numvalues) and
       (quit = false) do
begin
 if (temp.answers[I].CF <> 0) and
     (humans.onehuman[1].answers[I].CF <> 0)
 then
 begin
 outoforder := false;
 for J := I + 1 to ESres.numvalues do
     if temp.answers[I].rank >
```

```
            temp.answers[J].rank then
            outoforder := true;
      if outoforder = false then
            match := match + 1
      else
            quit := true;
   end
   else
      quit := true;
   I := I + 1;
end; {while}

allcount := 0;
for I := 1 to
              humans.onehuman[1].numvalues - 1 do
begin
 count := 0;
 J := I + 1;
 while J <= humans.onehuman[1].numvalues do
 begin
    if temp.answers[I].rank >
      temp.answers[J].rank then
      count := count + 1;
    J := J + 1;
 end;
 allcount := allcount + count;
end;

tau := 1 - (4 * allcount) /
        (humans.onehuman[1].numvalues *
        (humans.onehuman[1].numvalues - 1));
```

At this point in the program, the information on the results from both the expert system and the human expert(s) has been collected and analyzed. As seen in the above segment of code, this information is then sorted based on the certainty factor of the answers. After the information is in the proper order, the actual tabulating of Kendall's tau begins.

The next segment of code shows how the processing needed to calculate the distance comparability:

```
FILL_IN_LISTS (ESres, humans);
EScount := 0;
for I := 1 to ESres.numvalues do
     EScount := EScount + ESres.answers[I].CF;
humancount := 0;
for I := 1 to humans.onehuman[1].numvalues do
     humancount := humancount +
             humans.onehuman[1].answers[I].CF;

if EScount > humancount then
     distance := sqrt( (EScount * EScount) -
          (humancount * humancount))
else
     distance := sqrt( (humancount *
     humancount) - (EScount * EScount));
```

Again, at this point of the program, the results from the expert system and human expert(s) have already been collected and analyzed. The procedure FILL_IN_LISTS balances the list of recommendations for the expert system and human expert(s). If a possible solution does not appear on one of the lists, that solution is included with a certainty factor of zero. Balanced lists are needed to calculate both order compatibility and distance comparability.

In addition to the measures of order compatibility and distance comparability, the answers of the expert system and the answers of the human expert(s) are displayed in the report generated by the execution analysis subsystem. A sample report of the executionanalysis information is shown in Table 7.2.

2. Coverage Analysis

The coverage analysis subsystem evaluates the effectiveness of test cases in an execution suite with respect to its coverage of the rules within the knowledge base and the input and output equivalence classes. The execution information accumulated and maintained by the validation test manager and an analysis of the knowledge base are utilized in the determination of this coverage. The subsystem data flow diagram for the coverage analysis subsystem within the validation analyzer and enhancer is shown in Figure 7.19.

Table 7.2
Execution Analysis Report

EXECUTION ANALYSIS

Performance Comparison

Test Case Number	Total Order	Partial Order	Kendall's tau	Distance Compar.
1	N	Y/1	1.00	82.61
2	Y	Y/2	1.00	156.12

Actual Results

Test Case 1

Human Expert Answers	Expert System Answers
valpolicella 85	valpolicella 80
burgundy 30	

Test Case 2

Human Expert Answers	Expert System Answers
gewuerztraminer 65	gewuerztraminer 100
chardonnay 100	chardonnay 100

a. Coverage of Knowledge Base

The information generated by the coverage analysis subsystem includes the effectiveness of test cases in an execution suite in validating the rules within the knowledge base. A rule is validated if that rule is fired in the execution of a test case. The rules fired, the coverage index, and the effectiveness index are ascertained for each test case in an execution suite. The coverage index is the percentage of rules within the knowledge base which a test case validated. The effectiveness index is the percentage of rules within the knowledge base which were validated exclusively by a test case. These indices are very effective in helping to determine the most useful test cases in order to build a comprehensive, yet manageable set of test cases. An example of the tabular report produced about this information is shown in Table 7.3.

b. Coverage of Equivalence Classes

The effectiveness of the coverage over the input and output equivalence classes is also calculated in the coverage analysis subsystem. Input and output equivalence classes, which are de-

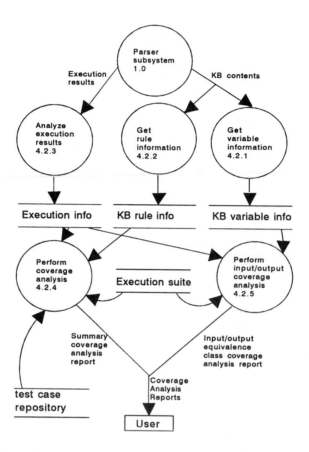

FIGURE 7.19.   Subsystem Data Flow Diagram for Coverage Analysis Facility within Validation Analyzer and Enhancer Component

scribed further in the next section on test case suggestions, are techniques for partitioning the information domain of a software product so as to comprehensively validate all aspects of the product with a minimum number of test cases. The effectiveness of coverage over these partitions by the execution suite is analyzed in the coverage analysis subsystem.

The data needed to produce the report on coverage of equivalence classes comes from three sources: the M.1 knowledge base, the execution suite, and the execution results. From the M.1 knowledge base, the names of all the input and output variables and their possible values, which were obtained earlier from the analysis of the M.1 knowledge base, are utilized. The values of all the variables in the test cases in the execution and the M.1 results

Table 7.3
Summary Coverage Report

COVERAGE ANALYSIS

Summary Information

Number of Test Cases in Execution Suite: 4
Number of Test Cases in Repository: 6
Number of Knowledge Base Rules: 56
Overall Coverage Index: 0.55

Test Case Number	Coverage Index	Effectiveness Index
1	0.13	0.04
2	0.34	0.20
3	0.20	0.05
4	0.14	0.09

for each test case are also used to build this report. The following segment of code shows the extracting of the needed data, the analysis of the data and the building of this report:

```
infilename := kbfilename[1] + kbfilename[2] +
              kbfilename[3] + kbfilename[4] +
              'ivar.val';
assign (infile, infilename);
reset (infile);
numinputs := 0;
while not eof (infile) do
begin
 numinputs := numinputs + 1;
 for I := 1 to maxword do
    read(infile,
        inputvars[numinputs].variablename[I]);
 readln(infile);
 readln(infile,
        inputvars[numinputs].numvalues);
```

```
for J := 1 to
        inputvars[numinputs].numvalues do
begin
for I := 1 to maxword do
    read(infile, inputvars
     [numinputs].possiblevalues[J,I]);
readln(infile);
inputvars[numinputs].numtimesused[J] := 0;
    for I := 1 to maxtestcases do
    inputvars[numinputs].inwhichtc[J,I]
            := false;
 end; {for J}
end; {while not eof}
close (infile);
assign (infile, 'runcases.val');
reset (infile);
readln(infile, numtcs);
for I := 1 to maxvalues do
begin
 outputclass[I].name := blank25;
 outputclass[I].numtimesused := 0;
 for J := 1 to maxtestcases do
     outputclass[I].inwhichtc[J] := false;
end; {for I}
for I := 1 to numtcs do
begin
 readln(infile, numvarvalues);
 for J := 1 to numvarvalues do
 begin
 for K := 1 to maxword do
        read(infile, tempvar[K]);
 for K := 1 to maxword do
        read(infile, tempval[K]);
 readln(infile);
 for K := 1 to numinputs do
 begin
    if inputvars[K].variablename = tempvar
    then begin
    for L := 1 to inputvars[K].numvalues do
    begin
    if tempval =
```

```
                inputvars[K].possiblevalues[L] then
                begin
                inputvars[K].numtimesused[L] :=
                inputvars[K].numtimesused[L] + 1;
                inputvars[K].inwhichtc[L,I] := true;
            end; {if values matched}
            end; {for L}
          end; {if variablenames matched}
          end; {for K}
  end; {for J}
  readln(infile, tempnum);
  readln(infile, numouts);
  for L := 1 to tempnum do
  begin
    for K := 1 to numouts do
  begin
    for M := 1 to maxword do
        read (infile, temp25[M]);
      readln(infile, tempCF);
      outputclass[K].name := temp25;
  end; {for K}
  end; {for L}
  end; {for I}
  close (infile);

  assign (infile, 'perfout.val');
  reset (infile);
  readln(infile, numtcs);
  for I := 1 to numtcs do
  begin
   readln(infile, tempnum);
   for J := 1 to tempnum do
   begin
      read(infile, ch);
      K := 1;
      while (ch <> ' ') do
      begin
        temp25[K] := ch;
        K := K + 1;
        read(infile, ch);
   end; {while}
```

```pascal
for K := K to maxword do
      temp25[K] := ' ';
readln(infile, tempCF);
for L := 1 to numouts do
begin
  if outputclass[L].name = temp25 then
  begin
   outputclass[L].numtimesused :=
      outputclass[L].numtimesused + 1;
   outputclass[L].inwhichtc[I] := true;
         end; {if tempCF}
  end; {for L}
 end; {for J}
end; {for I}
close (infile);


assign (infile, 'inputcov.val');
rewrite(infile);
writeln(infile,'C O V E R A G E      A N A L Y S I S');
writeln(infile);
writeln(infile);
writeln(infile,'INPUT EQUIVALENCE CLASSES');
writeln(infile,'========================');
writeln(infile);
writeln(infile);
writeln(infile);
writeln(infile, 'Equivalence Classes
            Coverage Index  Test Cases Where Used');
writeln(infile, '———————————————
                ——————————  ————————————————');
for I := 1 to numinputs do
begin
 writeln(infile);
 writeln(infile, inputvars[I].variablename);
 for J := 1 to inputvars[I].numvalues do
 begin
 write(infile, '   ',
        inputvars[I].possiblevalues[J],
        inputvars[I].numtimesused[J] /
        numtcs:9:2, '                    ');
 for K := 1 to numtcs do
```

```
    if inputvars[I].inwhichtc[J,K] = true then
        write (infile, K:4);
  writeln (infile);
  end;
end; {for I}
close(infile);
assign (infile, 'outputco.val');
rewrite(infile);
writeln(infile,'C O V E R A G E        A N A L Y S I S');
writeln(infile);
writeln(infile);
writeln(infile,'OUTPUT EQUIVALENCE CLASSES');
writeln(infile,'=========================');
writeln(infile);
writeln(infile);
writeln(infile);
writeln(infile, 'Equivalence Classes              Coverage Index
Test Cases Where Used');
writeln(infile, '————————————             ————————
————————————');
for I := 1 to numouts do
begin
  writeln(infile);
  write(infile, outputclass[I].name,
        outputclass[I].numtimesused /
        numtcs:9:2, '                        ');
  for K := 1 to numtcs do
  if outputclass[I].inwhichtc[K] = true then
      write (infile, K:4);
  writeln (infile);
end; {for I}
close(infile);
```

The reports produced by this segment of code in the coverage analysis subsystem are shown in Tables 7.4 and 7.5.

3. Test Case Suggestions

The test case suggestions subsystem analyzes the effectiveness of the validation coverage of the test cases along with the contents of the knowledge base. The result of these analyses is the generation of test case suggestions which, if applied with the current

Table 7.4
Coverage Analysis Report on Input
Equivalence Classes

COVERAGE ANALYSIS

INPUT EQUIVALENCE CLASSES

Equivalence Classes	Coverage Index	Test Cases Where Used
main-component		
meat	40	1, 3
poultry	20	2
fish	40	4
unknown	20	5
has-sauce		
yes	40	1, 5
no	60	2, 3, 4
sauce		
cream	20	5
spicy	0	
sweet	20	1
tomato	0	
tastiness		
average	60	1, 3, 4
delicate	20	5
strong	20	2
preferred-body		
full	40	2, 3
light	60	1, 4, 5
medium	0	
preferred-color		
red	40	1, 3
white	60	2, 4, 5
unknown	0	
preferred-sweetness		
dry	0	
medium	80	2, 3, 4, 5
sweet	20	1
unknown	0	

execution suite, will provide more complete validation of the knowledge base. The subsystem data flow diagram for this subsystem within SAVES is shown in Figure 7.20.

Table 7.5
Coverage Analysis Report on Output
Equivalence Classes

COVERAGE ANALYSIS

OUTPUT EQUIVALENCE CLASSES

Equivalence Classes	Coverage Index	Test Cases Where Used
burgundy	20	3
chablis	20	2
chenin-blanc	20	4
chardonnay	0	
gamay	0	
gewuerztraminer	0	
pinot-noir	20	1
riesling	0	
sauvignon-blanc	0	
soave	0	
valpolicella	0	
zinfandel	20	5

Techniques and tools for producing a representative set of test cases have been utilized in the development of conventional software, as discussed in Section V of Chapter 3, but have only

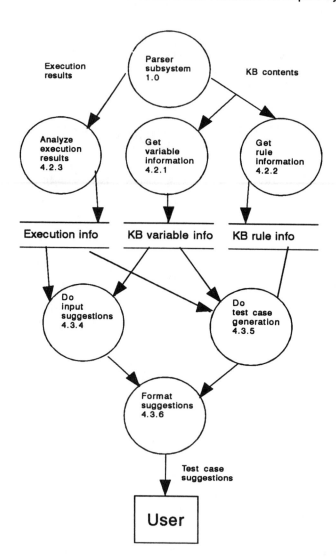

FIGURE 7.20. Subsystem Data Flow Diagram for Building Test Case Suggestions with Validation Analyzer and Enhancer Component

been suggested in the development of expert systems. The list of criteria, presented in Section V of Chapter 6, is an example of suggestions on how to develop test cases which effectively validate an expert system. The objective of such suggestions is the development of a comprehensive, yet manageable set of test cases.

The design of a representative set of test cases is needed for the validation of expert systems. Exhaustive testing of all possible input combinations for any software system is impractical. Equivalence partitioning is a technique used in black box testing to develop test cases. As discussed in Sections V of Chapter 3 and of Chapter 6, black box testing is the testing approach used in validation for both conventional software and expert systems. The functional requirements of a software product serve as the basis for deriving test cases for black box testing. In input equivalence partitioning, the input domain of a software product is "partitioned into a finite number of equivalence classes such that a test of a representative of each class will, by induction, test the entire class, and, hence, the equivalent of exhaustive testing of the input domain can be performed." [Goodenough, 1975] Guidelines for defining equivalence classes are:

1. If an input condition specifies a range, one valid and two invalid equivalence classes are defined.
2. If an input condition requires a specific value, one valid and two invalid equivalence classes are defined.
3. If an input condition specifies a member of a set, one valid equivalence class and one invalid equivalence class are defined.
4. If an input condition is Boolean, one valid class and one invalid class are defined. [Pressman, 1987]

In addition to equivalence classes for the input of a software product, equivalence classes for the output are also beneficial in devising a comprehensive set of test cases for the validation process. [Sommerville, 1989] Output equivalence classes are defined similarly to input equivalence classes.

The concepts of input and output equivalence partitioning are utilized by the validation analyzer and enhancer component of SAVES. The input and output equivalence classes equate to the input and output values required of the expert system. Although such partitioning is simplistic, for the size of knowledge base used in the evaluation of SAVES, this level of partitioning is sufficient. For larger, more complicated knowledge bases, an extended method of equivalence partitioning needs to be adopted.

For input equivalence classes, the test case suggestions subsystem is designed to analyze the coverage of these partitions by the test cases within the execution suite. Test case suggestions are then provided for the input equivalence classes which are inadequately validated. These test case suggestions are indications of the types of input values which need to be included in test cases if more complete validation is to be achieved.

The test case suggestions subsystem also generates test case suggestions for output equivalence classes which are inadequately validated. The output equivalence classes, not validated by the execution suite of test cases, are determined. For each unvalidated output equivalence class, a test case is generated by the validation analyzer and enhancer component of SAVES.

Generation of these test cases is based on a recursive, backward trace of the knowledge base. The trace begins with a knowledge base rule in which the consequent contains a member of the unvalidated output equivalence class. The trace, utilizing the analysis of the knowledge base, follows the antecedents of the rule backward until an antecedent, which contains only variables needed for test case input, is encountered. The first search through the knowledge base is for antecedent values which have a 100% certainty factor. If no value with 100% certainty factor exists, a second backward trace is undertaken which searches for the next highest certainty factor. The following segment of code shows a part of this recursive trace:

```
found := false;
I := 1;
while (I <= numrules) and (not found) do
begin
 J := 1;
 while (J <= rules[I].numberconclusions) do
 begin
 if rules[I].conclusions[J].variablename =
     apremisevariablename then
 begin
    if rules[I].conclusions[J].value =
        apremisevalue then
    begin
     for K := 1 to rules[I].numberpremises do
     begin
       isinput := false;
       for L := 1 to acase.numintestcase do
       begin
          if acase.varname[L] =
```

```
            rules[I].premises[K].variablename
      then
      begin
       isinput := true;
       loc := L;
      end; {if rules}
end; {for L}
if isinput = true then
      begin
      found := true;
      if rules[I].notinpremise[K] = false
      then
       acase.value[loc] :=
           rules[I].premises[K].value;
end {if}
else
begin
      temp :=
           rules[I].premises[K].variablename;
      X := 1;
      while (temp[X] <> ' ') and
            (temp[X] <> '-') do
             X := X + 1;
      if temp[X] = '-' then
      begin
       temp2 := 'preferred              ;
       Y := 10;
       while (temp[X]<>' ') and (X<=maxword)
       do begin
       temp2[Y] := temp[X];
       Y := Y + 1;
       X := X + 1;
       end;
       isinput := false;
       for L := 1 to acase.numintestcase do
       begin
          if temp2 = acase.varname[L] then
          begin
           isinput := true;
           loc := L;
```

```
            end; {if}
            end; {for L}
            if isinput = true then
                acase.value[loc] :=
                                rules[I].premises[K].value;
            end; {if temp[x] = '-'}

        {recursive call}
        BUILD(rules[I].premises[K].variablename,
                rules[I].premises[K].value,
                acase, rules, numrules, found);

            if found = false then
                BUILD2
                (rules[I].premises[K].variablename,
                rules[I].premises[K].value,
                acase, rules, numrules, found);
            end; {else}
            end; {for K}
          end; {if rules.value = apremisevalue}
      end; {if rules.variablename =
            apremisevariablename}
  J := J + 1;
  end; {J <= rules.numberconclusions}
  I := I + 1;
end; {while I <= numrules and not found}
```

This trace continues until all the antecedents of the rule containing a member of the unvalidated output equivalence class have been processed. The resulting test case is then analyzed for its completeness and its uniqueness. Missing values are supplied from the value range of an unspecified variable. Test cases, which are redundant of existing test cases, are discarded. The following segment of code in the test case suggestions subsystem within the validation analyzer and enhancer component of SAVES performs these tasks described above:

```
{get information about the variables and
values in the knowledge base which was
obtained earlier by parser subsystem}
```

```
infilename := kbfilename[1] + kbfilename[2] +
              kbfilename[3] + kbfilename[4] +
              'ivar.val';
assign (infile, infilename);
reset (infile);
I := 1;
while not eof (infile) do
begin
 for K := 1 to maxword do
     read (infile, temp[K]);
 atestcase.varname[I] := temp;
 readln (infile);
 readln (infile, numvals);
 for J := 1 to numvals do
     readln(infile);
 I := I + 1;
end; {while not eof}
close (infile);

{get information about test cases
in execution suite}
atestcase.numintestcase := I - 1;
for I := 1 to goals.numvalues do
begin
 if goals.firedcount[I] = 0 then
 begin
 for J := 1 to numrules do
 begin
    if rules[J].hasgoal = true then
    begin
     isequal := false;
     for K := 1 to
         rules[J].numberconclusions do
     begin
     if goals.values[I] =
         rules[J].conclusions[K].value then
         isequal := true;
     end; {for K}
     if isequal = true then
     begin
```

```
for K := 1 to
        rules[J].numberconclusions do
        atestcase.ESanswer[K] :=
              rules[J].conclusions[K].value;
atestcase.numESanswers :=
              rules[J].numberconclusions;
for L := 1 to maxtestvariables do
        atestcase.value[L] := blank25;
for L := 1 to
              rules[J].numberpremises do
begin
   BUILD (
   rules[J].premises[L].variablename,
   rules[J].premises[L].value,
   atestcase, rules, numrules, found);
end; {for L}
for L := 1 to
              atestcase.numintestcase do
begin
   if atestcase.value[L] = blank25 then
   begin
   if (atestcase.varname[L,1] = 'h') and
        (atestcase.varname[L,2] = 'a') and
        (atestcase.varname[L,3] = 's') and
        (atestcase.varname[L,4] = '-') then
   begin
   atestcase.value[L] := 'no                         ';
   end
   else
   begin
   atestcase.value[L] := 'unknown            ';
   end;
   end; {if atestcase.value}
end; {for L}
VERIFYTC (atestcase, isokay,
                    'allbuilt.val');
if isokay then
     STOREINFILE (atestcase,
                         'allbuilt.val');
end; {if isequal = true}
```

```
    end; {if rules.hasgoal = true}
  end; {for J}
  end; {if goals.firedcount = 0}
end; {for I}
```

These test cases are provided to the validation personnel as suggestions for more complete validation of the knowledge base. The validation personnel, e.g., the domain expert, need to analyze these test cases to determine their legitimacy. The domain expert also should provide the answer(s) for each test case which is to be added to the repository of validation test cases in SAVES.

III. EVALUATION OF SAVES

The evaluation of the demonstration prototype SAVES centers on the generation of test case suggestions. Aspects of the coverage analysis are discussed in relation to the effectiveness of the generated test cases. The effectiveness in validating the knowledge base and in validating the previously unvalidated input and output equivalence classes are examined in this evaluation.

Table 7.6 shows the three test cases which were initially entered into the SAVES repository and executed in the validation test manager component. These three test cases were randomly created. The overall coverage index for these three test cases is 46% of the knowledge base. The test case suggestions based on input equivalence classes are the values 'spicy' and 'sweet' for the input variable 'sauce', 'strong' for 'tastiness', 'fish' for 'main-component', 'yes' for 'has-veal', 'yes' for 'has-turkey', and 'medium' for 'preferred-sweetness'. The test case suggestions based on output equivalence classes are shown in Table 7.7. Many of the test case suggestions based on the input equivalence classes are also reflected in the test case suggestions based on the output equivalence classes. The three test cases based on the output equivalence classes are unique even though the last two differ by only one value. These test case suggestions are indicative of the coverage analysis on output equivalence classes which shows three output classes as unvalidated. 'Gamay', 'gewuerztraminer' and 'pinot-noir' are these three classes.

The first suggested test case is entered into the validation repository of SAVES. All four test cases (the three previous test cases

Table 7.6
Three Initial Test Cases

Test Case 1

main component	meat
has-veal	no
has-turkey	no
has-sauce	yes
sauce	tomato
tastiness	delicate
preferred-color	red
preferred-body	medium
preferred-sweetness	sweet

Test Case 2

main component	poultry
has-veal	no
has-turkey	no
has-sauce	yes
sauce	cream
tastiness	average
preferred-color	white
preferred-body	full
preferred-sweetness	dry

Test Case 3

main component	poultry
has-veal	no
has-turkey	no
has-sauce	no
tastiness	delicate
preferred-color	white
preferred-body	light
preferred-sweetness	sweet

and the new test case) are executed in the validation test manager component. The overall coverage index increased to 55% for these four test cases. The test case suggestions based on input equivalence classes are the value 'sweet' for the input variable 'sauce', 'strong' for 'tastiness', 'yes' for 'has-veal', 'yes' for 'has-turkey', and 'medium' for 'preferred-sweetness'. The input values, handled by the fourth test case, are no longer included in these test case suggestions. The test case suggestions based on output equivalence classes are the same as test cases 2 and 3 in Table 7.7 which were generated in the last execution. The coverage analysis for the output equivalence classes is also reflective of the two unvalidated equivalence classes. The coverage analysis shows the output equivalence classes of 'gamay' and 'pinot-noir' to be unvalidated.

Table 7.7
Three Generated Test Cases

Test Case 1

main component	fish
has-veal	no
has-turkey	no
has-sauce	yes
sauce	spicy
tastiness	unknown
preferred-color	white
preferred-body	full
preferred-sweetness	unknown

Test Case 2

main component	unknown
has-veal	no
has-turkey	no
has-sauce	yes
sauce	sweet
tastiness	average
preferred-color	red
preferred-body	medium
preferred-sweetness	sweet

Test Case 3

main component	unknown
has-veal	no
has-turkey	no
has-sauce	yes
sauce	sweet
tastiness	average
preferred-color	red
preferred-body	medium
preferred-sweetness	medium

The second test case from Table 7.7 is entered into the validation repository. All five test cases are executed in the validation test manager. The overall coverage index shows that these five test cases validated 70% of the knowledge base. The test suggestions based on the input equivalence classes are the value 'strong' for the input variable 'tastiness', 'yes' for 'has-veal', 'yes' for 'has-turkey', and 'medium' for 'preferred-sweetness'. There are no test case suggestions based on the output equivalence classes. The coverage analysis of the output equivalence classes shows that all classes have been validated and that test case 5 validated both the 'gamay' and 'pinot-noir' output equivalence classes.

This simple evaluation reveals marked improvement in the validation coverage of the knowledge base through the use of generated test cases. The randomly created test cases, which were first executed, validated 46% of the knowledge base, but with the addition of two SAVES generated test cases, the coverage increased to 70%. The input and output equivalence partitioning techniques are effectively utilized in SAVES for the generation of test case suggestions. The test case suggestions based on the output equivalence classes often include the suggestions proposed from the input equivalence classes. These features within SAVES provide an automated means to achieving more comprehensive validation of expert systems.

IV. CONCLUSIONS

SAVES (Suzanne and Abe's Validator for Expert Systems) is designed to show the feasibility of the recommendations for the techniques and tools needed for validating rule-based expert systems which are presented in this book. SAVES demonstrates that the maintenance of a repository of test cases, execution of test cases, application of statistical measures in analyzing the recommendations of an expert system against the recommendations of human expert(s), record-keeping and report-generating of previous validation testing, analysis of the effectiveness of test cases in the validation process, and generation of test case suggestions for more complete validation of the expert system can be automated and used effectively in the validation of rule-based expert systems. Such a set of techniques and tools is integral to providing effectual validation of rule-based expert systems.

SAVES supports validation activities throughout the development process and operational life of expert systems. SAVES is also designed to be generic; that is, it is not dependent on the development tool M.1 but can be adapted to any development tool or expert system shell which supports rule-based expert systems. All these features are desirable in a validation tool for expert system development. SAVES not only demonstrates the feasibility of therecommendations of this book but also the potential for commercial validation tools.

8 CONCLUSIONS

8   CONCLUSIONS

I. FUTURE RESEARCH TOPICS

The potential for effectual verification and validation of rule-based expert systems is exhibited in SAVES and the recommendations of this book. However, investigation into expert system verification and validation is in its infancy. We introduce several areas of research which are open to continued investigation. These areas are the generation of test case suggestions, coverage analysis techniques, complexity of the knowledge base, effect of different representational paradigms, incorporation of verification and validation into one development environment, tools for static and dynamic verification, and application of verification and validation concepts to neural networks.

The automated generation of test case suggestions to assist in the verification and validation of rule-based expert systems is still open to more research. Different measures may be useful in determining when to generate test cases. For example, the highest certainty factor for the members of an output equivalency class may be a measure for test case generation.

Coverage analysis in the verification and validation of rule-based expert systems advances further investigation. The usability of the effectiveness index and coverage index is one area of research. For example, utilizing these indices as indicators of redundant test cases, i.e., ones that verify or validate the same rules, is an interesting concept. Investigation into other partitioning techniques upon which to base coverage analysis is another area of research.

The effect of knowledge base complexity and the effect of different representational paradigms are interesting areas of research for expert system verification and validation. Verification and validation of rule-based expert systems is the focus of this book. The knowledge base used in the demonstration prototype SAVES is a relatively simple one. The application of the presented concepts of verification and validation to more complex knowledge bases requires investigation.

The examination of other representational paradigms, such as frames or semantic nets, is another area for future research. Ideally, the components of SAVES are applicable to any type of knowledge representation. The only components which are knowledge base specific are the knowledge base parsers utilized within SAVES. However, the question remains if these different representational paradigms create new problems in the verification and validation process which require extensions of the recommendations of this book.

The concepts of modularization and abstraction are being explored for use in the development of knowledge bases for rule-based expert systems. In these knowledge bases, the rules are grouped into sets and subsets and arranged in a tree-like structure. The effect of such structured rule-based systems on the verification and validation process is an open research issue. For example, the determination of input equivalence classes and output equivalence classes for structured knowledge bases is more complex than the approach presented in the demonstration prototype.

A related advancement in the development of expert systems is the use of an object-oriented approach to knowledge representation. Object-oriented expert systems utilize the concepts of encapsulation, inheritance, and message passing. Verification and validation of object-oriented expert systems presents new considerations for the coverage analysis aspects presented in this book. Multiple effectiveness indices and coverage indices may be required to assess the verification and validation of several layers in a knowledge base. Coverage of the objects and coverage within a subsystem related to one object are two examples of added complexity in the verification and validation of object-oriented expert systems.

An important area of research is the incorporation of verification and validation tools into one development environment. One limitation on the advancement of verification and validation for expert systems is the fragmentation of effort on these quality assurance techniques for expert systems. Small pockets of research, primarily into the verification of expert systems, are currently being undertaken in universities and industry. None of these efforts have attempted to incorporate the aspects of verification and validation into one development environment. Figure 7.1 is an example of the type of environment which needs to be examined.

Static and dynamic verification are open research issues for expert systems. Static verification tools such as EVA [Stachowitz, 1987], ARC [Nguyen, 1987], and VALIDATOR [Jafar, 1989] have addressed only simple rule-based expert systems. The verification of frame-based expert systems or object-oriented expert systems advances many unaddressed problems. Furthermore, the development of tools to assist in the dynamic verification of expert systems is an area of unexplored research. Concepts within SAVES may have applicability to dynamic verification.

The application of the verification and validation concepts presented in this book to the verification and validation of neural networks provides a possible research issue. Work on determining the equivalency between expert systems and neural networks [Kuncicky, 1990] has led to this consideration. If research shows that neural networks are equivalent to rule-based expert systems, then research into verifying and validating neural networks using the recommended verification and validation techniques and tools is a feasible project.

As seen in the preceding discussion, verification and validation of expert systems is open to many other avenues of research. The recommendations presented here are just the inception of understanding and advancement in the verification and validation of rule-based expert systems. The continuing demand for quality assurance techniques, such as verification and validation, in the development of expert systems will require continued research in these areas.

II. SUMMARY

The use of software quality assurance techniques such as verification and validation in the development of expert systems is essential for the widespread acceptance of expert system technology: "for any 'serious' system on which lives, health, money, or even happiness, are to depend it must be possible to verify and validate the answers that the system will give in any particular set of circumstances." [Oakley, 1989] The commercial computing community is demanding the same level of software quality assurance in expert systems as has been demanded in conventional software. Currently however, there exist no systematic methods, techniques, or tools in expert system technology for determining the quality of expert systems.

The contributions achieved in the recommendations of this book provide a comprehensive and systematic methodology for the verification and validation of rule-based expert systems. These contributions are the definition and demarcation of verification and validation, the specification of necessary testing activities, the delineation and definition of needed testing documentation, and the collection and usage of test cases.

A complete set of techniques and tools for the verification and validation of rule-based expert systems is established in this book. Objectivity in the verification and validation process is recommended and supported by the use of statistical measures in the comparison of expert system performance and human expert performance. The use of automated tools in the verification and validation of rule-based expert systems is suggested in the recommendations of this book. The feasibility of such automation is evidenced in the demonstration prototype SAVES. SAVES demonstrates the automation of the maintenance of a test case repository, execution of test cases, application of statistical measures, record-keeping and report-generating of previous testing, analysis of the effectiveness of test cases in the verification and validation process, and generation of test case suggestions for more complete verification and validation of the expert system.

The recommendations presented in this book adapt the concepts of verification and validation for conventional software. Concepts such as input equivalence partitioning and output equivalence partitioning are established techniques of conventional software which are applied in this book to the verification and validation of rule-based expert systems. The weaknesses and difficulties of current verification and validation and the lessons learned from past verification and validation experience are also addressed in the recommendations presented in this book.

The need for preliminary planning activities in the verification and validation process is satisfied through the development of the Project V & V Plan and test plans. The subjectivity and time-consuming, tedious tasks of verification and validation are eliminated through the use of statistical measures and automated tools. Finally, the recommendations in this book include new techniques or tools for verification and validation where the uniqueness of expert systems demands them and where current methods and practices are inadequate. Coverage analysis of the knowledge base and the automated generation of test cases for more comprehensive verification and validation of the knowledge base are examples of these techniques and tools.

The recommendations in this book promote a systematic, comprehensive approach to the verification and validation of expert systems. From the software crisis, we learned that haphazard testing of a few test cases cannot measure the quality of a software product. Nevertheless, the current approach to the verification and validation of expert systems uses such haphazard testing to assess expert system quality. Without evidence of the measure of quality, widespread acceptance of expert system technology by the commercial computing community will be delayed. A systematic, comprehensive methodology for verification and validation is necessary for demonstrating the measure of quality of expert systems.

APPENDICES

APPENDIX A

VERIFICATION AND VALIDATION TECHNIQUES AND TOOLS

Algorithm analysis
Analytical modeling
Assertion generation
Assertion processing
Cause-effect graphing
Code auditor
Comparator
Control-flow analyzer
Criticality analysis
Cross-reference generator
Database analyzer
Dataflow analyzer
Design-compliance analyzer
Execution-time estimator
Formal review
Formal verification
Functional testing
Inspections
Interactive test aids
Interface checker
Metrics
Mutation analysis
Program-description-language processor
Peer review
Physical-unit testing

183

Regression testing
Round-off analysis
Simulations
Sizing
Software monitors
Specification base
Structural testing
Symbolic execution
Test drivers
Test-coverage analyzer
Test-data generator
Test-support facilities
Timing
Tracing
Walkthroughs

APPENDIX B

CRITERIA	DEFINITION AND ACTIVITIES
Accuracy	how well a simulation reflects reality. Compare inferences made by rules with historic (known) data, observe correctness of the outcome.
Adaptability	possibilities for future development and application. Keep I/O and control rules general; revise facts and rules when new information is available. Periodically review the desirability of integrating with existing or proposed hardware or software systems. Should the system be self-modifying or context sensitive? Can it be customized for particular user needs?
Adequacy	the fraction of pertinent empirical observations that can be simulated. Establish list of parameters (variables, conditions, and relations) that influence inference outcome, determine which to include in rule set.

Appeal

usability; how well the knowledge base matches out intuition and simulates thought; practicability. Appeal is a potentially key criterion for marketability; test usability by assessing I/O friendliness relatively early in the development process. Test simulation and practicability on site in beta-development stage.

Availability

existence of other, simpler, validated knowledge bases that solve the same problem(s), important for determining eventual marketability. Will users perceive the need for a new rule-based system if other tools are already available and meet their needs?

Breadth

proportional to the number of rules used in the knowledge base. Determine the number of contexts within which the system should be expected to perform, and thus the number of pertinent parameters to account for in the rule set.

Depth

proportional to the number and kinds of variables chosen to describe each component in the model. Determine the range of conditions the system will address and which parameters are necessary to diagnose, classify, and/or advise for each condition. Depth will in turn determine necessary input data and user interface.

Face Validity

model credibility. Have knowledge base, inference structure, and output reviewed by credible human experts during early development of prototype and later expansion of full-scale system. Compile and report results.

Generality

capability of a knowledge base to be used with a broad range of similar problems. Define the general contexts within which the system can be expected to perform at expert levels and provide strong caution that use beyond these contexts may not yield accurate results.

Precision

capability of a model to replicate particular system parameters; also the number of significant figures used in numeric variables and computations. Ensure that all pertinent variations of parameters are represented in the rule base and facts. Express numbers as floating point or real format as necessary; use double precision for calculations, especially those involving matrix or linear algebra calculations.

Realism

accounting for relevant variables and relations. Establish parameters and functions in the rule base in the same terms and with the same conceptual models used by experts or end user audience. Realism is particularly important when developing the full-scale knowledge base, and also involves the logical order with which queries are made.

Reliability

the fraction of model predictions that are empirically correct (actually, part of a complex statistical analysis of the accuracy and correctness of the entire rule base). Reliability includes conditional and posterior probabilitiesof statistical utility of the likelihoods in the rules and outputs.

Resolution

the number of parameters of a system the model attempts to mimic. Identify which parameters need to be defined and represented in detail and which can be grouped into more general conditions or ignored.

Robustness

conclusions that are not particularly sensitive to model structure. Determine which input parameters are least and most significant in the form of the interim (diagnosis, classification) and final (advice, alarm) results and output. Be sure the latter are well defined in the rules and functions.

Sensitivity

the degree to which variations of knowledge base parameters induce outputs that match historical data. Specifically determine sensitivity of results to each input parameter by varying the parameter incrementally, holding all other parameters constant and matching model output with historical (known) data.

Technical and

identification and importance of all Operational Validity divergence in model assumptions from perceived reality. Carefully explicate the contexts, conditions, and assumptions that underlie the rules and relations. Discuss how each assumption limits model results. How do they affect model accuracy, reliability, robustness, and generality?

Turing Test

assessing the validity of a knowledge base by having human evaluators distinguish between the model's conclusions on a specific problem and a human expert's conclusions solving the same problem.

Usefulness

validates that the system contains necessary and adequate parameters and relationships for use in problem-solving contexts (if at least some model predictions are empirically correct). Usefulness is trivial for a full-scale system but important for prototyping and adding onto existing rule sets.

Validity   a knowledge base's capability of producing empirically correct predictions. Given the contexts within which the system is expected to operate well, determine how many actual conditions the system can accurately diagnose, classify, and advise. Determine the level of correctness with human experts in the same area and set realistic objectives for correctness of the knowledge base.

Wholeness   the number of processes and interactions reflected in the model. How complex is the rule base? How many factors does it use? Consider wholeness in light of adaptability.

REFERENCES

REFERENCES

[Agarwal, 1990] Agarwal, R. and Tanniru, M., "Systems Development Life-Cycle for Expert Systems" in Knowledge-Based Systems, 3 (3), 170–180, September 1990.

[Blackman, 1990] Blackman, M. J., "CASE for Expert Systems" in AI Expert, 5 (2), 26–31, February 1990.

[Boehm, 1978] Boehm, B., Brown, J. R., Kaspar, H., Lipow, M., MacLeod, G. J., and Merrit, M. J., Characteristics of Software Quality, North-Holland Publishing Company, New York, New York, 1978.

[Boehm, 1984a] Boehm, B. W., "Verifying and Validating Software Requirements and Design Specifications," IEEE Software, 1 (1), 75–88, January 1984.

[Boehm, 1984b] Boehm, B. W., "Software Life Cycle Factors," in Handbook of Software Engineering, Vick, C. R. and Ramamoorthy, C. V., Eds., Van Nostrand Reinhold Company, New York, New York, 1984, 494–518.

[Boehm, 1988] Boehm, B. W., "A Spiral Model of Software Development and Enhancement," IEEE Computer, 21, 61–72, May 1988.

[Chapnick, 1990] Chapnick, P., "Software Quality Assurance," AI Expert, 5 (2), 5–6, February 1990.

[Cho, 1987] Cho, C., Quality Programming: Developing and Testing Software with Statistical Quality Control, John Wiley & Sons, Inc., New York, New York, 1987.

[Cooper, 1979] Cooper, J. D. and Fisher, M. J., Eds., Software Quality Management, Petrocelli Books, Inc., New York, New York, 1979.

[Denning, 1986] Denning, P. J., "Towards a Science of Expert Systems," IEEE Expert, 1 (2), 80–83, Summer 1986.

[Deutsch, 1988] Deutsch, M. S. and Willis, R. W., <u>Software Quality Engineering: A Total Technical and Management Approach</u>, Prentice Hall, Englewood Cliffs, New Jersey, 1988.

[Dunn, 1990] Dunn, R. H., <u>Software Quality</u>, Prentice Hall, Englewood Cliffs, New Jersey, 1990.

[Eliason, 1990] Eliason, A. L., <u>Systems Development: Analysis, Design, and Implementation</u>, Scott, Foresman and Company, Glenview, Illinois, 1990.

[Evans, 1987] Evans, M. W. and Marciniak, J. J., <u>Software Quality Assurance and Management</u>, John Wiley & Sons, New York, New York, 1987.

[Fairley, 1985] Fairley, R. E., <u>Software Engineering Concepts</u>, McGraw-Hill Book Company, New York, New York, 1985.

[Fox, 1990] Fox, M. S., "AI and Expert System Myths, Legends, and Facts" in <u>IEEE Expert</u>, 3 (1), 8–20, February 1990.

[Freeman, 1987] Freeman, P., <u>Software Perspectives</u>, Addison-Wesley Publishing Company, Reading, Massachusetts, 1987.

[Gaschnig, 1983] Gaschnig, J. et al., "Evaluation of Expert Systems: Issues and Case Studies" in <u>Building Expert Systems</u>, Hayes-Roth, F., Waterman, D. A., and Lenat, D. B., Eds., Addison-Wesley Publishing Company, Inc., Reading, Massachusetts, 1983, 241–280.

[Geissman, 1988] Geissman, J. R. and Schultz, R. D., "Verification and Validation of Expert Systems" in <u>AI Expert</u>, 3 (2), 26–33, February 1988.

[Giarratano, 1989] Giarratano, J. and Riley, G., <u>Expert Systems: Principles and Programming</u>, PWS-Kent Publishing Company, Boston, Massachusetts, 1989.

[Goodenough, 1975] Goodenough, J. B. and Gerhart, S. L., "Toward a Theory of Test Data Selection" in <u>IEEE Transactions on Software Engineering</u>, SE-1 (2), June 1975.

[Green, 1987] Green, C. J. R. and Keyes, M. M., "Verification and Validation of Expert Systems" in <u>Proceedings Western Conference on Expert Systems</u>, IEEE Computer Society Press, Piscataway, New Jersey, June 2–4, 1987, 38–43.

[Guida, 1989a] Guida, G. and Spampinato, L., "Assuring Adequacy of Expert Systems in Critical Application Domains: A Constructive Approach," in <u>The Reliability of Expert Systems</u>, Hollnagel, E., Ed., Ellis Horwood Limited, Chichester, England, 1989, 134–167.

[Guida, 1989b] Guida, G. and Tasso, C., "Building Expert Systems: From Life Cycle to Development Methodology," in <u>Topics in Expert System Design: Methodologies</u>

and Tools, Guida, G. and Tasso, C., Eds., Elsevier Science Publishers, North-Holland, Amsterdam, The Netherlands, 1989, 3–24.

[Gupta, 1991] Gupta, U. G., Ed., Validating and Verifying Knowledge-Based Systems, IEEE Computer Society Press, Los Alamitos,California, 1991.

[Hall, 1988] Hall, L. O., Friedman, M. and Kandel, A., "On the Validation and Testing of Fuzzy Expert Systems" in IEEE Transactions on Systems, Man, and Cybernetics, 18 (6), 1023–1027, November/December 1988.

[Harmon, 1985] Harmon, P. and King, D., Expert Systems: Artificial Intelligence in Business, John Wiley & Sons, Inc., New York, New York, 1985.

[Harmon, 1990] Harmon, P. and Sawyer, B., Creating Expert Systems for Business and Industry, John Wiley & Sons, Inc., New York, New York, 1990.

[Hetzel, 1988] Hetzel, W., The Complete Guide to Software Testing, Second Edition, QED Information Sciences, Inc., Wellesley, Massachusetts, 1988.

[Hollnagel, 1989] Hollnagel, E., "Evaluation of Expert Systems," in Topics in Expert System Design: Methodologies and Tools, Guida, G. and Tasso, C., Eds., Elsevier Science Publishers, North-Holland, Amsterdam, The Netherlands, 1989, 377–416.

[Hu, 1987] Hu, S. D., Expert Systems for Software Engineers and Managers, Chapman and Hall, New York, New York, 1987.

[IEEE, 1983] IEEE Standard Glossary of Software Engineering Terminology, IEEE Std. 729–1983, February 1983.

[Irgon, 1990] Irgon, A., Zolnowski, J., Murray, K. J., and Gersho, M., "Expert System Development: A Retrospective View of Five Systems" in IEEE Expert, 5 (3), 25–40, June 1990.

[Jafar, 1989] Jafar, M. J., "A Tool for Interactive Verification and Validation of Rule-Based Expert Systems," Ph.D. dissertation, Department of Systems and Industrial Engineering, University of Arizona, Tucson, Arizona, 1989.

[Kang, 1990] Kang, Y. and Bahill, A. T., "A Tool for Detecting Expert-System Errors" in AI Expert, 5 (2), 46–51, February 1990.

[Kiper, 1992] Kiper, J. D., "Structural Testing of Rule-Based Expert Systems" in ACM Transactions on Software Engineering and Methodology, 1 (2), 168–187, April 1992.

[Kuncicky, 1990] Kuncicky, D. C., Hruska, S. I., and Lacher, R.C., "Hybrid Systems: The Equivalence of Expert System and Neural Network Inference," submitted for publication.

[Lane, 1986] Lane, N.E., "Global Issues in Evaluation of Expert Systems" in Proceedings of the 1986 IEEE International Conference on Systems, Man, and Cybernetics, IEEE Computer Society Press, Piscataway, New Jersey, 1986, 121–125.

[Liebowitz, 1988] Liebowitz, J., An Introduction to Expert Systems, Mitchell Publishing, Inc., Santa Cruz, California, 1988.

[Liebowitz, 1986] Liebowitz, J., "Useful Approach for Evaluating Expert Systems" in Expert Systems, 3 (2), 86–92, April 1986.

[Marcot, 1987] Marcot, B., "Testing Your Knowledge Base" in AI Expert, 2 (8), 42–47, August 1987.

[McCabe, 1976] McCabe, T., "A Software Complexity Measure" in IEEE Transactions on Software Engineering, 2, 308–320, December 1976.

[McCall, 1977] McCall, J., Richards, P., and Walters, G., "Factors in Software Quality," 3 volumes, NTIS AD-A049–014, 015, 055, November 1977.

[Mills, 1987] Mills, H. D., Dyer, M., and Linger, R. C., "Cleanroom Software Engineering" in IEEE Software, 19–25, September 1987.

[Musa, 1989] Musa, J. D. and Ackerman, A. F., "Quantifying Software Validation: When to Stop Testing?" in IEEE Software, 19–27, May 1989.

[Nguyen, 1987] Nguyen, T., Perkins, W., Laffey, T. and Pecora, D., "Knowledge Base Verification" in AI Magazine, 69–75, Summer 1987.

[Oakley, 1989] Oakley, B., "Evaluation Criteria for Expert Systems," in Expert Systems in Production and Services II: From Assessment to Action?, Proceedings of the International Workshop on Expert Systems in Production and Services held in Chicago, Illinois, September 13–15, 1988, Bernold, T. and Hillenkamp, U., Eds., Elsevier Science Publishing Company, Inc., New York, New York, 1989, 123–129.

[O'Keefe, 1987] O'Keefe, R. M., Balci, O. and Smith, E. P., "Validating Expert System Performance" in IEEE Expert, 2 (4), 81–90, Winter 1987.

[O'Leary, 1987] O'Leary, D. E., "Validation of Expert Systems — With Applications to Auditing and Accounting Expert Systems" in Decision Sciences, 18 (3), 468–486, Summer 1987.

[O'Leary, 1990] O'Leary, T. J., Goul, M., Moffitt, K. E. and Radwan, A. E., "Validating Expert Systems" in IEEE Expert, 5 (3), 51–58, June 1990.

[Oliver, 1987] Oliver, A.E.M., "Techniques for Expert System Testing and Validation," Proceedings of Third International Expert Systems Conference, Learned Information, Oxford, England, June 1987, 271–276.

[Ould, 1986] Ould, M. A. and Unwin, C., Eds., Testing in Software Development, Cambridge University Press, Cambridge, England, 1986.

[Parsaye, 1988] Parsaye, K. "Acquiring and Verifying Knowledge Automatically" in AI Expert, 3 (5), 48–63, 1988.

[Parrington, 1989] Parrington, N. and Roper, M., Understanding Software Testing, Ellis Horwood Limited, Chichester, England, 1989.

[Powell, 1986] Powell, P. B., "Planning for Software Validation, Verification, and Testing," in Software Validation, Verification, Testing, and Documentation, Andriole, S. J., Ed., Petrocelli Books, Princeton, New Jersey, 1986, 3–78.

[Prerau, 1990] Prerau, D. S., Developing and Managing Expert Systems: Proven Techniques for Business and Industry, Addison-Wesley Publishing Company, Reading, Massachusetts, 1990.

[Pressman, 1987] Pressman, R. S., Software Engineering A Practitioner's Approach, McGraw-Hill Book Company, New York, New York, 1987.

[Price, 1990] Price, C., "Improving Present-day Toolkits," in Knowledge Engineering Toolkits, Price, C.J., Ed., Ellis Horwood, New York, New York, 1990, 75–89.

[Ramamoorthy, 1987] Ramamoorthy, C.V., Shekhar, S., and Garg, V., "Software Development Support for AI Programs" in Computer, 20 (1), 30–40, January 1987.

[Royce, 1970] Royce, W.W., "Managing the Development of Large Software Systems: Concepts and Techniques" in Proceedings of the IEEE WESCON, (held in Los Angeles, CA, August 25–28, 1970), IEEE Press, New York, New York, August 1970, 1–9.

[Rushby, 1988] Rushby, J., Quality Measures and Assurance for AI Software, NASA Contractor Report 4187, National Aeronautics and Space Administration, Scientific and Technical Information Division, 1988.

[Sarmiento, 1989] Sarmiento, C. D., "Applying Traditional Software Engineering Methodologies to the Development of Expert Systems: What Works and What Doesn't?" in Advances in Artificial Intelligence Research, Volume 1, Fishman, M. B., Ed., JAI Press Inc., Greenwich, Connecticut, 1989, 349–360.

[Schach, 1990] Schach, S. R., Software Engineering, Aksen Associates, Inc., Boston, Massachusetts, 1990.

[Snedecor, 1989] Snedecor, G. W. and Cochran, W. G., Statistical Methods, Eighth Edition, Iowa State University Press, Ames, Iowa, 1989.

[Sommerville, 1989] Sommerville, I., Software Engineering, Addison-Wesley Publishing Company, Reading, Massachusetts, 1989.

[Stachowitz, 1987] Stachowitz, R. A. and Combs, J. B., "Validation of Expert Systems" in <u>Proceedings of the Twentieth Hawaii International Conference on System Sciences 1987</u> (Volume 1: Architecture, Decision Support Systems and Knowledge-Based Systems), Western Periodicals Co., North Hollywood, California, 686–695.

[St. Johanser, 1986] St. Johanser, J.T. and Harbridge, R.M., "Validating Expert Systems: Problems & Solutions in Practice" in <u>Knowledge Based Systems '86</u>, Proceedings of the International Conference held in London, July 1986, Online Publications, London, England, 1986, 215–229.

[Taylor, 1989] Taylor, J. R., "Control of Software Reliability," in <u>The Reliability of Expert Systems</u>, Hollnagel, E., Ed., Ellis Horwood Limited, Chichester, England, 1989, 37–63.

[Tuthill, 1990] Tuthill, G. S., <u>Knowledge Engineering: Concepts and Practices for Knowledge-Based Systems</u>, TAB Professional and Reference Books, Blue Ridge Summit, Pennsylvania, 1990.

[Wallace, 1989] Wallace, D. R. and Fujii, R. U., "Software Verification and Validation: An Overview" in <u>IEEE Software</u>, 10–17, May 1989.

[Walters, 1979] Walters, G. F., "Application of Metrics to a Software Quality Management (QM) Program," in <u>Software Quality Management</u>, Cooper, J. D. and Fisher, M. J., Eds., Petrocelli Books, Inc., New York, New York, 1979.

[Waterman, 1986] Waterman, D., <u>A Guide to Expert Systems</u>, Addison-Wesley Publishing Company, Reading, Massachusetts, 1986.

[Webster, 1987] <u>Webster's New World Dictionary</u>, Simon & Schuster, Inc., New York, New York, 1987.

[Wilson, 1990] Wilson, D., "First Case Study: Expert System Shell," in <u>Knowledge Engineering Toolkits</u>, Price, C. J., Ed., Ellis Horwood, New York, New York, 1990, 135–142.

[Zeide, 1990] Zeide, J. S. and Liebowitz, J., "A Critical Review of Legal Issues in Artificial Intelligence," in <u>Managing Artificial Intelligence and Expert Systems</u>, DeSalvo, D. A. and Liebowitz, J., Eds., Yourdon Press, Englewood Cliffs, New Jersey, 1990.

INDEX

INDEX